Managing Maritime Safety

Shipping is a pillar of global trade, with 90 per cent of the world's trade in goods and raw materials carried by ship. Despite the economic benefits this delivers, maritime operations can be dangerous, and when accidents occur the consequences are serious. Consequential outcomes from hazards at sea include serious injury, death, loss of cargo and destruction of the marine environment.

Managing Maritime Safety will give you a thorough understanding of contemporary maritime safety and its management. It provides varying viewpoints on traditional safety topics in conjunction with critical discussions of the international safety management code and its application. The book also offers new perspectives on maritime safety such as ship and equipment design for safety and the relevance of safety management systems, in particular the application of the International Safety Management code to remote controlled or autonomous ships. The authors all work in the maritime industry, as practitioners, in education, research, government and classification. The combination of wide-ranging and extensive experience provides an unprecedented span of views with a strong connection to the real issues in the maritime domain.

This book sets out to provide much needed consolidated knowledge for university level students on maritime safety management, incorporating theoretical, historical, research, operational and design perspectives.

Helle A. Oltedal holds a PhD in safety management and safety culture within the Norwegian-controlled shipping industry. She is currently Program Manager of the Maritime Safety (MarSafe) Research Program at Western Norway University of Applied Sciences. Her research interests include aspects of organizational safety management and safety culture within the maritime industry.

Margareta Lützhöft is a master mariner and holds a PhD in human–machine interaction on the ship's bridge. Presently she holds a position as Professor of Maritime Human Factors in the Maritime Safety (MarSafe) Research Program at Western Norway University of Applied Sciences. Her research interests include human-centred design and the effects of new technology, and she has published in these and other areas relating to maritime safety.

'The world in general and the maritime industry specifically are facing an unprecedented time of change. Terms like Industry 4.0, IoT, and automation abound and are set against a world with growing population, environmental challenges and that glorious goal of "sustainability". Can the maritime industry manage this change while remaining safe or even improving safety? This book takes the essential move of stepping back to look at the "big picture", identifying essential lessons before trying to redefine the value of humans in the future of maritime transport. Our industry can be polarised on the impact of adopting new technology; however, change is coming and coming fast. It is my hope that this book will help us identify how humans can best use technology, improve design, deal with increased uncertainty and that as an industry we can embrace the future with our eyes wide open.'

— David J Patraiko, MBA, FNI; Master Mariner;
Director of Projects, The Nautical Institute

Managing Maritime Safety

Edited by Helle A. Oltedal and
Margareta Lützhöft

Routledge
Taylor & Francis Group

LONDON AND NEW YORK

First published 2018
by Routledge
2 Park Square, Milton Park, Abingdon, Oxon OX14 4RN

and by Routledge
711 Third Avenue, New York, NY 10017

Routledge is an imprint of the Taylor & Francis Group, an informa business

© 2018 selection and editorial matter, Helle A. Oltedal and Margareta Lützhöft; individual chapters, the contributors

The right of Helle A. Oltedal and Margareta Lützhöft to be identified as the authors of the editorial material, and of the authors for their individual chapters, has been asserted in accordance with sections 77 and 78 of the Copyright, Designs and Patents Act 1988.

British Library Cataloguing-in-Publication Data
A catalogue record for this book is available from the British Library

Library of Congress Cataloging-in-Publication Data
Names: Oltedal, Helle, editor of compilation. | Lutzhoft, Margareta, editor of compilation.
Title: Managing maritime safety / edited by Helle Oltedal and Margareta Lutzhoft.
Description: Milton Park, Abingdon ; New York, NY : Routledge, 2018. | Includes bibliographical references and index.
Identifiers: LCCN 2017039808| ISBN 9781138559226 (hbk) | ISBN 9781138559233 (pbk) | ISBN 9780203712979 (ebk)
Subjects: LCSH: Shipping--Security measures. | Merchant marine--Security measures. | Navigation--Safety measures.
Classification: LCC VK203 .M33 2018 | DDC 363.12/3--dc23
LC record available at https://lccn.loc.gov/2017039808

ISBN: 978-1-138-55922-6 (hbk)
ISBN: 978-1-138-55923-3 (pbk)
ISBN: 978-0-203-71297-9 (ebk)

Typeset in Times New Roman
by HWA Text and Data Management, London

Contents

Figures

Tables

Contributors

Chad Allen has been involved with the marine industry for twenty-five years, working both at sea and ashore for a large international container line and more recently working for a trade association representing international shipping in Canada. He holds a Diploma in Nautical Science from the Marine Institute, St Johns, Canada, a Bachelor of Maritime Studies degree from Memorial University of Newfoundland, Canada, and is currently working towards a Master of Maritime Management degree at Memorial University. He holds a masters intermediate voyage license issued by Transport Canada.

Bjørn-Morten Batalden is an Associate Professor in Nautical Science and Deputy Head of Department at UiT the Arctic University of Norway. He is a master mariner and holds a PhD in risk management and societal safety from the University of Stavanger, Norway, studying safety management practices in the offshore support vessel segment. Prior to his PhD, he worked several years as the Safety Manager and Company Security Officer for a ship management company in Norway. His current research focuses mainly on how the maritime industry adapts to new technologies and the move towards autonomous maritime operations.

Jonathan V. Earthy is Human Factors Coordinator and technical authority for Lloyd's Register Marine & Offshore, based in the Technical Policy Group. He has worked for Lloyd's Register for twenty-five years. He has a BSc from Durham University, UK and a PhD from Aston University, UK. Prior to Lloyd's Register he worked for BP Exploration. His research speciality is the assurance of the quality of human factors integration and human-centred design. His other research interests are: the introduction of human factors into the marine industry, standardization of human factors, sociotechnical systems, and system resilience. He contributes to a wide range of advisory, steering and standards committees.

Michelle R. Grech is Vessel Operations Manager at the Australian Maritime Safety Authority (AMSA) responsible for the delivery of strategic and operational safety outcomes across the Australian maritime industry. She obtained her PhD from the University of Queensland, Australia, in human factors, specializing in fatigue, workload and situation awareness and has published extensively in these areas. She is currently leading international efforts to support fatigue risk management approaches in shipping.

Jon Ivar Håvold is a professor in organizational and work psychology at the Norwegian University of Science and Technology (NTNU) in Ålesund, Norway. He obtained his PhD from NTNU in Trondheim within maritime safety culture. He has been teaching management, cultural understanding and safety management to maritime students for around twenty years. His main research interest is organizational safety management and safety culture within the maritime industry.

Margareta Lützhöft is a master mariner, trained at Kalmar Maritime Academy in Sweden. Presently she holds a position as Professor of Maritime Human Factors at the Maritime Safety (MarSafe) Research Program at the Institute of Nautical Studies, Western Norway University of Applied Sciences. Her research interests include human-centred design and the effects of new technology, and she has published in these and other areas relating to maritime safety.

Helle A. Oltedal is Program Manager of the Maritime Safety (MarSafe) Research Program at the Institute of Nautical Studies, Western Norway University of Applied Sciences. She holds a PhD in risk management and societal safety from the University of Stavanger in Norway, specializing in safety management and safety culture within the maritime industry. Her research interests address organizational, operational and interpersonal aspects of how maritime risk and safety is managed.

James Parsons started his career in the Canadian Arctic and worked globally with multinational officers and crews. He has sailed on very large crude carriers, liquefied petroleum gas carriers, product carriers, general cargo, container, seismic, gaff rigged schooners, and cruise ships. Since 1997 he has been teaching at the Marine Institute, St Johns, Canada, and is involved in the development and delivery of training courses and programmes for various global clients. Most recently, he has a keen interest in Arctic shipping research. He holds a master mariner foreign going license and a PhD in marine transportation economics; his PhD was focused on Arctic shipping. Jim is Canada's civilian expert to NATO's Transport Group Ocean Shipping and an active volunteer with the Master Mariners of Canada.

Viet Dung Vu is a PhD student at the Australian Maritime College. Starting as a seafarer, Viet Dung later decided to follow an academic career and joined Vietnam Maritime University as a lecturer. He began his research at the Australian Maritime College in 2016 and currently works to improve the inclusion of human factors considerations in the design of marine navigation equipment.

Acknowledgements

We thankfully acknowledge MARKOM 2020 for funding the project.

Our greatest appreciation to Elizabeth Clouter for reviewing and Paula Rice for proofreading.

We also thank our friends and colleagues for reviewing.

Thank you Pieter Six for checking our accidents, and Quan Zhang for being a fatigue expert.

Eternal gratitude to Vu – the emperor of references.

Introduction

The maritime industry is characterized as high risk. A maritime accident may have catastrophic consequences for us and our society. Systematic safety management began in the early twentieth century and other high-risk industries such as nuclear and aviation have been the subjects of extensive research and are often used as references in safety management literature. All contributing authors to this book work within the maritime industry in a combination of areas: seafaring, lecturing, training and research. We all identified a need for an academic book on safety management addressing the maritime industry in particular. The characteristics of the maritime industry, sailing the seven seas away from modern society, emergency services and rescue personnel, means that the challenges the maritime industry faces are different. Thus, based on these needs, we have brought together a group of maritime experts to fill the gap in safety literature. We invite the reader to browse or read cover to cover depending on your interests. To assist in your reading, we provide a glimpse into the chapters below.

In Chapter 1, Helle A. Oltedal sets the stage for maritime safety management. The concept of safety is examined and the development of systematic safety management by following the five ages of safety is introduced. It elaborates on the influence of focal factors such as technology, behaviours and human error, socio-technical issues and culture and the concept of resilience, as well as how developments are reflected within the maritime industry. The chapter ends by proposing the age of interaction and adaptive safety models.

In Chapter 2, James Parsons and Chad Allen start with the history of safety management from a legal perspective. We go on a journey from Plutarch to the Polar code, making stops along the way to follow the changes in approaches to safety until we arrive at what today is called modern safety management. We have a look at navies and the Titanic and their influence on the development of this. We visit Plimsoll and Jones, to see their impact on maritime safety. The 20th century brings us from prescriptive regulation in the shape of SOLAS, MARPOL, STCW, Collision and the Load Line Convention to the advent of a new look as heralded by the promising introduction of a Just Culture perspective.

Chapter 3 brings us to the central theme, the safety management system, by Bjørn-Morten Batalden and Helle A. Oltedal. We follow how the International Safety Management code ISM was developed and implemented in shipping. The chapter brings home the need for this regulation by reviewing numerous accidents, and discusses the expected and unexpected effects of the ISM. The benefits and drawbacks of self-regulation are unpacked, with a discussion of possible differing interpretations of this topic on both ship and shore. The latter half of the chapter delves into the processes for organizational learning, information analysis and the implementation into decisions.

The varying views on culture, including safety culture, are then expanded upon in Chapter 4, 'Culture and maritime safety', by Jon Ivar Håvold and Helle A. Oltedal. The concept of culture has a long and opaque history and can be used at many levels for groups of people or organizations. The IMO has introduced instruments to create a safety culture and this chapter shows examples from shipping that work as well as examples of conflict.

Chapter 5, 'The human contribution', by Helle A. Oltedal and Margareta Lützhöft, puts a positive spin on the role of humans in the maritime system. We argue that much can be done to reap the benefits of having humans in place, while avoiding the drawbacks. The changing views of causality are discussed, as are some of the challenges of avoiding 'human error'. The effectiveness of initiatives to address the issues, such as legislation or BRM, are analysed and suggestions for improvements are made.

After this, we expand on the perception of risk and safety from the human perspective in Chapter 6, 'Risk perception', by Michelle R. Grech. This chapter examines risk as the centrepiece of safety and explores how risk and safety are perceived in the maritime industry and how that influences behaviour. We look at human biases and limitations regarding our ability to judge risks objectively as well as the consequences, and we discuss what effects that may have on human behaviour. Furthermore, we examine how this could influence intervention strategies and the safety climate in an organization.

Given what we now know about humans and their capabilities and limitations, we take a fresh look at what we can do to support them. We move away from traditional approaches such as training, warning and punishing in Chapter 7, 'Design for safety', by Margareta Lützhöft and Viet Dung Vu. This is a review of legal and other instruments to increase safety in the hardware of shipping. There is a detailed presentation of initiatives already taken, and a look at the added value of human-centred ships and technologies. How do we produce safe operational conditions and make it difficult to make mistakes while showing clearly what can/ should be done?

Chapter 8 by Jonathan V. Earthy and Margareta Lützhöft is 'hot off the press', and very pre-regulation. We open up a discussion on some of the issues concerning remote-controlled or even autonomous ships, seeing that this is a rapidly developing sector in engineering, but perhaps less so in non-technical and humanistic fields. We look at maritime regulation and what may need to be done, at SMS and how that fits, and consider how the autonomous system might be safely manned. Finally, we explore the requirements for varying methods of control.

We hope that you learn from this book as we have learned from writing it.

1 Setting the stage for maritime safety management

Helle A. Oltedal

Introduction

'Never before have so few done so much for so many'. When opening the Year of the Seafarer in 2010, these words, a quotation paraphrased from one of Winston Churchill's most famous speeches, were strikingly declared by Efthimios E. Mitropoulos, Secretary General of the International Maritime Organization (IMO) (Oltedal, 2011). The profession of seafaring and shipping is crucial in our society. The worldwide population of seafarers serving in international trade is estimated to be approximately 1,545,000 people, representing virtually every nationality. Worldwide, there are about 90,917 vessels, registered in over 150 nations, which carry about 90 per cent of the world's trade; thus, these more than one and a half million seafarers are transporting goods for the benefit of the world's 7.49 billion people (United Nations, 2016). The seafarers' and shipping industry's global importance is commonly highlighted by the phrase 'without shipping, half the world would starve and the other half would freeze'.

Seafaring is seen as a high-risk occupation, and due to the nature of seafaring, the risk factors differ from those found at a traditional on-shore workplace. Some of the characteristic stressors and risk factors are known to be long-term separation from home and family, social isolation, long working hours along with high workload and reduction in number of crew members, lack of shore leave, fatigue, high levels of work-related stress, accidents and maritime disasters, exposure to hazardous substances, dangers from piracy, and environmental stressors such as ship motion, noise and vibrations which may impact work performance and thus risk and safety (Slišković & Penezić, 2015).

When something goes wrong – and at some time all organizational systems will experience unwanted events – the maritime transport system is also distinct with respect to handling and investigating such events. When on the ocean, rescue personnel, fire squads, medical assistance and the like are not as easily available as for shore-based industries. Moreover, when accidents happen on land, there are normally crash scenes with bits and pieces that may support understanding of the chain of events leading up to the accident. Within the maritime sector, the accident scene may very well be nothing but water, with the vessel in bits and pieces somewhere on the seabed, kilometres below the surface.

In maritime safety research, human and social aspects have gained less attention than the technical ones (Hetherington, Flin, & Mearns, 2006), even though the human element is said to be the major causal factor in unwanted events (Butt, Johnson, Pike, Pryce-Roberts, & Vigar, 2013). However, it is gratifying to recognize that this kind of research has been gradually increasing over recent decades.

This book intends to present a critical review of contemporary maritime safety research, and this chapter will discuss the safety concept itself along with the historical development of safety theories and safety models in order to set the stage. While this first chapter presents and discusses the historical development of safety theories and models at a general level, the following chapters aim to present in more detail the state of the art and challenges in safety management at organizational and operative levels, the importance of safety culture, risk perception and the human contribution. When managing safety, making strategic decisions and developing new safety measures, it is crucial that one has a basic understanding of which fundamental rationale the theories and models of use are founded on as well as their limitations, to ensure appropriate understanding and application. Thus, in the following. safety management will be discussed on a conceptual level, providing the reader with a brief tour of the history, and how theoretical and practical limitations and challenges are driving further developments, starting with a presentation of three attributes around which the safety management system should be designed.

Three attributes of safety management

To manage safety in any domain, it is essential to gain an understanding of how and why accidents and incidents occur in the first place and what makes an organization safe. There are various theoretical models of accident and incident investigation, with different explanations for sources of risk and safety, which provide guiding principles and constraints for the development of safety actions and an overall safety management system.

It has been suggested that safety management in any domain should be designed around three attributes of organizations and their environments (Grote, 2012), which are:

1   The kinds of safety to be managed: process versus personal safety.
2   The general approach to managing uncertainty as a hallmark of organizations that manage safety: minimizing uncertainty versus coping with uncertainty.
3   The regulatory regime within which safety is managed: external regulation versus self-regulation.

These three attributes, and their relevance to the domain of managing maritime safety, will be discussed in depth in Chapter 3 and briefly addressed below.

It has been pointed out that the choice of theoretical approach towards safety management should depend upon the kind of safety to be managed – distinguishing between process safety and personal safety – due to differences in the visibility and complexity of risks (Grote, 2012). In process safety, the risks and uncertainties to be managed are directly linked to the primary work task of the organization, such as transporting goods and people. Potential damage results from failures in the execution of processes linked to this task, such as the processes of designing the vessel, maintenance or navigational failure. Breaches of process safety do not necessarily cause harm to the human operators involved. For instance, in May 2015 the cruise vessel *Hamburg* ran aground off Scotland and suffered considerable raking damage to the hull. The accident caused no injuries to passengers or crew (Marine Accident Investigation Branch [MAIB], 2016b). However, when running aground there is always a potential for larger-scale accidents, and that potential is a feature which distinguishes process accidents from personal accidents.

In personal safety, on the other hand, potential damage always concerns the human operator(s), but is not necessarily directly linked to the primary work task. This could be a seafarer tripping on a ladder or falling into the cargo hold. Throughout this book, we refer mostly to process safety.

Another reason why it is important to distinguish process safety from personal safety is the use of key performance indicators. Within shipping – and other high-risk industries – it is common to use key performance indicators that are related to personal safety as an indicator for both personal and process safety, such as lost working days due to injury or number of reported personal accidents, non-conformities or near misses. We have not found any maritime accident reports clearly discussing this, but it is known to have been an underlying problem both in the BP Texas City Refinery disaster in March 2005 where explosions and fires killed 15 people and injured another 180 (US Chemical Safety and Hazard Investigation Board, 2007), and the Deepwater Horizon Incident in April 2008 where 11 people were killed (Deepwater Horizon Study Group, 2011). Both organizations had a person-oriented approach towards safety, at the cost of process safety, and did not pay sufficient attention to underlying strategic organizational decisions, training, risk assessment or risk management. Both organizations were operated by BP, and both incident reports point to a poor organizational safety culture, which is further elaborated in Chapter 4 of this book.

The two quotes below are taken from the two incident reports, pinpointing the problem:

> One underlying cause was that BP used inadequate methods to measure safety conditions at Texas City. For instance, a very low personal injury rate at Texas City gave BP a misleading indicator of process safety performance. In addition, while most attention was focused on the injury rate, the overall safety culture and process safety management (PSM) program had serious deficiencies.
> (US Chemical Safety and Hazard Investigation Board, 2007, p. 19)

> The Deepwater Horizon had an outstanding record of preventing lost-time incidents. In 2008, the Deepwater Horizon had received an award for its safety record, and on the day of the explosion there was a ceremony on board the rig celebrating seven years without a lost-time incident.
> (Deepwater Horizon Study Group, 2011, p. 38)

The second attribute to be considered when managing safety is related to the general approach organizations have towards risk and safety, with the distinction between minimizing versus coping with uncertainty. When minimizing uncertainty, the organization to a large degree relies on central planning, a high degree of standardization and little operative freedom. The shipping industry is known to be highly proceduralized (Oltedal, 2011). For instance, the investigation into the *Hoegh Osaka* incident – a car and truck carrier that developed a severe list and stranded in January 2015 – did address the extended use of detailed procedures and checklists (MAIB, 2016a). In this case the five checklists used for discharging and loading combined a total of 213 tick boxes. Several of the critical items were ticked off, but not completed, which highlights the problem with this approach – the possibility for critical work operations becoming lost among a large number of minor tasks.

Proceduralization and standardization is also in stark contrast to the expertise that is required to handle the uncertainty that is found in maritime high-risk operations. Dreyfus and Dreyfus (1986) argue that procedures and checklists are more helpful in guiding the actions of those that are under training and new to the work tasks at hand, and less for those that are considered as competent professionals. It is even argued that extensive use of rules such as procedures and checklists will hinder the development of the expertise needed to handle new complex situations, which often are characterized by a large degree of uncertainty (Dreyfus, Dreyfus, & Athanasiou, 2000). Such situations often require a different strategy which Grote (2012) calls coping with uncertainty.

However, the challenge is to balance these strategies, as both are needed, depending upon the part of the operation that is performed. Perrow (1999) suggests categorizing organizational systems into two groups, of either *linear* or *complex* interactions.

> *Linear interactions* are those in expected and familiar production or maintenance sequence, and those that are quite visible even if unplanned.
> *Complex interactions* are those with unfamiliar sequences, or unplanned and unexpected sequences, which are either not visible or not immediately comprehensible.

We agree with Perrow (1999) that systems with complex interactions are those that are most prone to system accidents, which might be understood as equal to Grote's (2012) term process accidents, but we argue that maritime operations are both, and the balance between linearity and complexity depends upon the operation that is performed. For instance, when sailing an anchor handling vessel to a location, the operation may be considered more linear than complex, but during the actual anchor handling, the complexity and uncertainty increases. There are still isolated tasks that might be considered as linear, but when seen in context the scene is different. Those who are particularly interested in how a standard operation may devolve into an uncontrolled complex situation are recommended to read the investigation report of the loss of the anchor handler *Bourbon Dolphin* in April 2007 (Norges offentlige utredninger, 2008).

Grote's (2012) third attribute, the regulatory regime within which safety is managed, distinguishes between external regulation and self-regulation. We argue that the maritime industry is regulated in both ways. The international regulations provided by the IMO through the International Safety Management (ISM) Code fosters self-regulations. However the maritime industry – in particular petroleum and gas related activities – are also externally regulated, by their customers. The regulatory mechanisms are further elaborated in Chapter 3, along with safety management in general.

In the following section, we will expand on the history of safety management in order to give the reader an understanding of the historical background as well as enabling the reader to evaluate within which approach of safety management a given company operates. We start with a discussion of the concept of safety itself.

Safety as a concept

Systematic examination of the causes of accidents began in the early 20th century, and its early roots are to be found in the 1931 Herbert W. Heinrich publication, *Industrial Accident Prevention: A Scientific Approach* (Heinrich, 1931). Based on his research, he proposed that 88 per cent of all workplace accidents were caused by unsafe acts by operators; however, he encouraged organizations to control hazards, not merely focus on human behaviour. Heinrich is also well known for proposing some of the earliest accident causation models – the Domino theory, which describes an accident as a chain of discrete events which occur in a particular temporal order, and the iceberg theory proposing that there is a proportional link between minor injuries and major accidents. Although Heinrich's research and theories have been criticized in more modern research (Manuele, 2002) he should still be acknowledged for being a pioneer, placing accident prevention and investigation in the spotlight.

Since then, the focus on safety, in theory as well as in practice, has traditionally been on the outcomes or products rather than on the processes. The common concern is with what safety achieves, rather than with what safety is. Safety is generally defined as the freedom from serious and unacceptable outcomes or in some cases the freedom from unacceptable risks (potential outcomes). In recent years this has become known as Safety-I. Since this focuses on something that we should be without, something that should be avoided, the 'obvious' solution is to try to do just that: to eliminate hazards, to prevent things from going wrong and to protect ourselves against the outcomes in case it happens anyway (Hollnagel, 2017). From this perspective, simple linear or complex linear models are often used to identify system weaknesses, which may be organizational, technical or non-technical, such as violations, poor communication, inadequate leadership or the like, followed by safety measures that address that particular weakness. Assessment of the overall state of the system is then measured by freedom from injuries, accidents, safety violations or near misses.

Understanding safety as being *without* something (accidents, incidents, mishaps, etc.) is not the only possibility. Safety can also be understood as being *with* something, namely acceptable outcomes. When an event takes place, when something is done, the outcome can for given conditions either be acceptable or unacceptable, but not both at the same time. This perspective – also called Safety-II – focuses on positive outcomes rather than negative, and defines safety as a condition where as much as possible goes well. This is the logical opposite to the understanding represented by Safety-I. The focus now is on something that the organization should be with, something that should be sought or attained. Thus, under this perspective, the approach is to try to understand how things go well, to support and facilitate that, and to ensure that things go right both in the short and the long term.

Expanding the focus of safety from the rare unacceptable outcomes to include also the common acceptable outcomes has a number of practical consequences. The understanding of how things happen must include non-linear models and explanations; the measurements must look at outcomes that increase as safety improves; learning must go beyond accidents and incidents to include typical, everyday performance; the management of safety must finally go beyond preventing unsuccessful outcomes and look at how to produce acceptable outcomes as well (Hollnagel, 2017). Instead of only investigating unwanted events, the Safety-II perspective would promote successful operations and scrutinize how people adapt to constantly changing and complex operations filled with uncertainty – and thus avoid an unwanted event.

The following section presents in more detail how the different accident and safety models have developed thorough the Safety-I perspective into the era of Safety-II and resilience engineering.

The historical development of safety management

Pillay (2015) describes the historical development of safety management, and identifies five ages of safety. The first age is closely associated with technology; the second with behaviours and human error; the third with socio-technical issues; the fourth with culture and the fifth with resilience. These five ages of safety may also be categorized into three eras. The predominant safety models in the eras and ages of safety are depicted in Table 1.1, and will be further elaborated below.

Table 1.1 The development of safety management

Eras of Safety	Ages of Safety	Predominant Safety Models
First Era	• Technological • Human factors	Linear models or sequential models
Second Era	• Socio-technical • Cultural	Complex linear models or epidemiological models
Third Era	• Resilience	Complex non-linear models or systemic models

The first era of safety – focus on technology and humans

The first *era* of safety includes the two ages where first technical aspects, followed by human aspects, were commonly pointed to as the causes of unwanted events. At the time, the predominant safety models were linear or sequential.

The first *age* of safety lasted from the beginning of the 19th century until after the Second World War and is labelled the technical age, where accidents were largely attributed to mechanical and structural failures, and these were thought to be preventable by following technical standards and guidelines issued by professional engineers, architects and designers (Pillay, 2015). In line with this – as a direct consequence of the *Titanic* disaster in 1912 – the International Convention for the Safety of Life at Sea (SOLAS Convention) was adopted. The SOLAS convention, which is still in force, regulates the technical issues of maritime safety such as improved design, vessel structure and equipment.

During this age, the predominant causes of accidents were explained through the domino theory by Heinrich and Granniss (1959) and other simple linear or sequential models. The *simple linear models* assume that accidents are the culmination of a series of events or circumstances which interact sequentially with each other in a linear way, and those accidents can be avoided by taking action and eliminating one of the causes in the linear sequence.

As a result, technical improvements did reduce the frequency, severity and consequences of shipping incidents. However, the reduction of failures in technology revealed the underlying influence of human error in accident causation; and in line with the second age of safety, such factors were addressed, although still using an approach that assumed unwanted events evolved in a linear and sequential manner.

In the period between the two world wars, research within personnel selection, training and motivation theories was increased and ushered in the second stage of safety in the 1960s and 1970s, which is labelled the *age* of human factors. The adoption of the Standards of Training, Certification and Watchkeeping (STCW) Convention in 1978 led to a shift towards the individual seafarer as a cause of accidents.

Low and Sua (2000) suggest that unsafe conditions are symptoms of mismanagement and that simple linear and sequential models fail to clarify the relationships between personal and organizational factors, and could therefore lead to wrong interpretations of the underlying factors that explained the cause of an accident. Brière, Chevalier, and Imbernon (2010) agree with the claim that the misinterpretation might be particularly common for high-ranking staff that usually do not work on site and lack in-depth safety knowledge. Other critics also claim that this can lead to an over-emphasis on worker behaviour and not enough attention on systems, and that large-scale organizational accidents and minor accidents differ systematically in many regards, as also pointed out by Grote (2012), when differentiating between process safety and personal safety.

Although both the technical approach through SOLAS and the focus on behaviour and human factors through STWC led to major safety improvements, several marine accidents revealed that human error could be induced by organizational factors and poor management standards. The capsizing of the ferry *Herald of Free Enterprise* in 1987 that resulted in 187 fatalities is seen as one of the major drivers for moving the maritime industry into the third and the fourth ages of safety through the adoption of the International Safety Management (ISM) code in 1994, and a shift in focus towards organizational causes, latent factors and safety culture. This acknowledgement led to the second era of safety, with a shift towards systemic and organizational factors.

The second era of safety – focus on systems and culture

The second *era* of safety includes the third and fourth ages, where socio-technical and cultural causes commonly were included as underlying factors in incidents – also referred to as root causes. At the time, the predominant safety models were complex linear or epidemiological and these are still predominant in accident investigation. The third *age* of safety was established in the 1990s moving from the technological and individual approach towards safety management systems and organizational factors (Hale & Hovden, 1998). It was realized that humans were rarely the sole cause of accidents or errors, and that human performance was based on complex interactions of the socio-technical system that constitutes an organization.

In the early 1980s, Perrow (1999) argued that technological advances made organizations and systems more prone to accidents, by making the systems not only tightly coupled but inherently complex, so much so that he claimed that accidents in such systems are unavoidable – coining the term 'Normal Accidents'. Perrow addresses maritime accidents specifically, and argues that it is characteristics of the system itself that cause maritime accidents. The system is to be understood not only in terms of technology, but how it is organized at national and international levels, the political game when developing regulations, the insurance system, production pressures, owner structure, ship register and flag state, as well as onboard organizational structure that interplay to let accidents happen. One example is the *Scandinavian Star* fire in April 1990, killing 159 people (Norges offentlige utredninger, 1991). The ship had only just been put in service after renovations, and it is claimed that she was not seaworthy when the voyage started. The crew members were new, and some did not understand either a Scandinavian language or English; they were fatigued and not properly familiarized with the vessel, her emergency equipment, the emergency plan and so on. All in all, the ship was not prepared for sea. The ship's captain tried unsuccessfully to cancel the voyage the day before. During the aftermath, it was unclear who owned the vessel and who had the legal responsibility, as a group of different owners, countries and shell companies were involved. In June 2014, the *Scandinavian Star* case was reopened, and 1 June 2017, 27 years after the accident, the Norwegian commission of enquiries released their most recent report.

The complex linear models assume that accidents are a result of a combination of unsafe acts and latent hazard conditions within the system following a linear path. The factors furthest away from the accident are attributed to actions of the organization or the environment, while the factors at the sharp end closest to the accident are attributed to humans in the operations. These models assume that accident prevention should focus on strengthening barriers and defences. The epidemiological models are a variation of the complex linear models which see events leading to accidents as analogous to the spreading of a disease. Unsafe acts made by operators at the sharp end are seen as symptoms of underlying latent organizational problems, which may lie dormant in the organization for years before they reveal themselves – the disease. One of the key researchers within this approach is James Reason and his theory of organizational accidents (Reason, 1997).

The concept of safety culture was also brought forward in this second era, claiming that risk and accidents were often induced by a poor safety culture, where organizations and operators did not have a culture that supported safe operations. Reason (1997) introduced the concept of safety culture and organizational culture in relation to how safety information is handled, or in the worst case ignored. The cultural perspective is further addressed in Chapter 5.

The simple and complex linear models, which are still the predominant models in use, have contributed to our understanding of accidents. However, it was argued that sequential, linear and epidemiological accident models are inadequate for capturing the dynamics and non-linear interactions between system components in complex socio-technical systems, and this led to the introduction of new accident models. These are based on systems theory and classified as systemic accident models. They endeavour to describe the characteristic performance on the level of the system as a whole, rather than on the level of specific cause–effect 'mechanisms' or even epidemiological factors (Qureshi, 2007). In contrast to the former approaches, systemic and complex non-linear models view accidents as emergent phenomena, which arise due to the complex and non-linear interactions among system components, which leads us to the third era of safety and the age of resilience.

The third era of safety – the age of resilience

Both the first and second era of safety belong to what has become the Safety-I perspective. The age of resilience represents a new approach towards accident modelling – the Safety-II perspective. The Safety-II perspective represents a new way of thinking, and the predominant safety models include complex non-linear models or systemic models. This view is distinctive from the Safety-I perspective, particularly in two ways. First, the idea of a linear cause–effect mechanism is discarded, and these new models focus on processes with feedback loops of information and control. Second, the models within the Safety-I perspective are based on hindsight, learning from previous unwanted events, errors and flaws at different levels in the organization or system. Compare this to complex non-linear models or systemic models which consider the performance of a system as a whole. Accidents are treated as the result of flawed processes involving interactions among people, social and organizational structures and engineering activities, and physical and software system components. The focus is on these processes, and how to create processes that are robust yet flexible to adapt to the complexity of the real world, and thus proactively avoid accidents by monitoring and controlling risk (Dekker, Hollnagel, Woods, & Cook, 2008).

Resilience engineering is still a theoretical concept, and we are not aware of any empirical studies supporting this perspective – at least not within the maritime sector – and the predominant safety models currently used all fall into the Safety-I perspective (Schröder-Hinrichs, Praetorius, Graziano, Kataria, & Baldauf, 2015). Today, when it comes to preventing major accidents, systemic accidents models or other more complex models for in depth analysis are mainly used for investigating technological, organizational and human elements, along with their interrelationships (Hovden, Albrechtsen, & Herrera, 2010).

Chapter 3 will address necessary changes to safety management systems to incorporate the resilience engineering perspective.

Managing maritime safety: where we stand today

During the twentieth century, attempting to understand the mechanisms of accident causation has produced many theories and models affecting the way people think about safety, and what factors they identify as risks. Some of the models are linear, suggesting that one factor leads to the next and the next leads to an accident, whereas some are non-linear and complex, covering many factors that in combination lead up to an accident.

Simple linear models have been criticized for not capturing the complexity in the real world. The idea of linear causality derives from natural science, where such correlations are found; for example, when you drop a stone it always falls. Within social sciences, which involve human beings, such causal relationships do not exist. The linear sequential models are still useful in situations that are approximate to linear causal relationship. For example, we know that humans need oxygen to breathe. Thus when entering an enclosed space, it is wise to either measure the oxygen level or use a breathing apparatus. The problem arises when such models are used in situations where causal relationships are more absent. In light of their limitations, the Safety-II perspective and resilience engineering offers an alternative way to understand safety – to study the ongoing processes such as planning, communication and cooperation which takes place ahead of any actions with a determined outcome, and thus gain understanding of how these processes support safe operations.

Still, causal accident models and the Safety-I perspective are predominant among practitioners, and safety measures have been developed to control risk factors revealed in a previous accident. As human beings we have an inherent fundamental need to be in *control.* When facing situations perceived as uncontrolled, we might attempt to create an 'illusion of control', which is the assumption of control when there is no true control over the situation or event (Leotti, Iyengar, & Ochsner, 2010). In an organizational context, control involves decisions, and sometimes change, to promote safe operations at a strategic as well as operative level. After working within the maritime industry for decades, it is our experience that control strategies that take the form of adding a new procedure or checklist are overemphasized. Such an approach might be a symptom of repetitive anxiety-avoidance (Reason, 1997). Anxiety-avoidance refers to an organization that discovers a technique for reducing collective anxiety of not being safe and repeats it over and over again regardless of its actual effectiveness. The following example illustrates how such an approach towards safety may undermine the attempt to manage safety.

Some companies, in particular those working for the oil and gas industry, have written procedures to ensure that a lid is always put on a coffee cup. The intention is to avoid human injury from hot coffee spill. Those observed without the lid are reported as non-conforming.

When using this approach uncritically, human actions and deviations are compared to a preproduced standard and found to be erroneous. In an attempt to gain control, new and even more detailed measurements may be developed, thereby creating a vicious cycle resulting from the anxiety of not being in control. In addition, despite the need for control being inherent in our human nature, we should realize and accept that as humans we are not capable of being in control of any future situations, as the future is dynamic, evolving here and now, and we should therefore prioritize carefully.

When deciding on safety measures, the possible consequences of a risk should be considered, and it may be decided that low-risk situations are left uncontrolled. With reference to the above example, an attempt to control hot coffee spill with procedures, along with other low-risk situations, may result in an extensive procedural system difficult to relate to in operations. Safety critical procedures may 'drown' or get lost in those that are perceived as less important for safety, as with the *Hoegh Osaka* incident.

Attempting to control minor risk also contributes to overload of information into the safety management system. All reports – as non-conformities – need resources to be processed in the safety management system. When resources are used on safety measures and processing to control minor risk, fewer resources are left available for handling safety. Also, when the focus is on issues that are perceived as less safety critical, the operator may lose trust in the overall safety management system, with perceptions like 'it is nothing other than loads of paperwork, reports, procedures and check-lists, while no one does anything about what matters'.

Many safety professionals and practitioners are not fully aware of the strengths and weaknesses of existing accident and safety models. In addition, accidents come in many sizes, shapes and forms and therefore one model or one type of universally applicable explanation is unrealistic. Some accidents are simple, and therefore only need simple explanations and simple models. Some accidents are complex, and need more advanced models and methods to be analysed and prevented. The development of accident models, from the domino model and onwards, has been forced by cases that defied the then current ways of thinking.

We have over time been faced with the need for models that can account for the variability of accidents, both analytically and for predictions; however, the introduction of a new model does not necessarily mean that those already existing become obsolete. It rather serves to highlight their strengths and weaknesses and thereby, in a sense, to ascertain when they should be used and when not.

With the introduction of resilience engineering, a new way of thinking about safety was introduced: from linear causality to non-linearity and from focus on unwanted outcome to wanted outcome. To prevent accidents, it is important to understand the processes leading up to a successful operation and understand how humans successfully adapt to a constantly changing environment, and then channel resources to enforce those processes.

Although the intention of resilience engineering is good, the wording may suggest associations to the technical field, which could work against it. In a system including human beings it is important to understand the processes of interaction between humans as well as the interaction between humans and technology. We therefore propose the introduction of a new more comprehensive description of the current age of safety: the age of 'Interaction', and the predominant model the 'Adaptive Safety Model'.

References

Brière, J., Chevalier, A., & Imbernon, E. (2010). Surveillance of fatal occupational injuries in France: 2002–2004. *American Journal of Industrial Medicine, 53*(11), 1109–18.

Butt, N., Johnson, D., Pike, K., Pryce-Roberts, N., & Vigar, N. (2013). *15 Years of Shipping Accidents: A Review for WWF*. Southampton: Southampton Solent University.

Deepwater Horizon Study Group (2011). *Final Report on the Investigation of the Macondo Well Blowout*. Retrieved from http://ccrm.berkeley.edu/pdfs_papers/bea_pdfs/dhsgfinalreport-march2011-tag.pdf

Dekker, S., Hollnagel, E., Woods, D., & Cook, R. (2008). *Resilience Engineering: New Directions for Measuring and Maintaining Safety in Complex Systems*. Lund: Lund University School of Aviation.

Dreyfus, H., & Dreyfus, S. (1986). *Mind Over Machine*. New York: Free Press.

Dreyfus, H., Dreyfus, S., & Athanasiou, T. (2000). *Mind Over Machine*. New York: Simon & Schuster.

Grote, G. (2012). Safety management in different high-risk domains: All the same? *Safety Science, 50*(10), 1983–92.

Hale, A. R., & Hovden, J. (1998). Management and culture: The third age of safety. A review of approaches to organizational aspects of safety, health and environment. *Occupational Injury: Risk, Prevention and Intervention*. Boca Raton, FL: CRC Press.

Heinrich, H. (1931). *Industrial Accident Prevention: A Scientific Approach*. New York: McGraw-Hill.

Heinrich, H., & Granniss, E. R. (1959). *Industrial Accident Prevention*. New York: McGraw-Hill Book Co.

Hetherington, C., Flin, R., & Mearns, K. (2006). Safety in shipping: The human element. *Journal of Safety Research, 37*(4), 401–11.

Hollnagel, E. (2017). *Safety-II in Practice: Developing the Resilience Potentials*. Abingdon: Routledge.

Hovden, J., Albrechtsen, E., & Herrera, I. A. (2010). Is there a need for new theories, models and approaches to occupational accident prevention? *Safety Science,* 48(8), 950–6.

Leotti, L. A., Iyengar, S. S., & Ochsner, K. N. (2010). Born to choose: The origins and value of the need for control. *Trends in Cognitive Sciences,* 14(10), 457–63.

Low, S. P., & Sua, C. S. (2000). The maintenance of construction safety: Riding on ISO 9000 quality management systems. *Journal of Quality in Maintenance Engineering,* 6(1), 28–44.

Manuele, F. A. (2002). *Heinrich Revisited: Truisms or Myths.* Ithaca, NY: National Safety Council Press.

Marine Accident Investigation Branch (2016a). *Report on the Investigation into the Listing, Flooding and Grounding of Hoegh Osaka – Bramble Bank, The Solent, UK on 3 January 2015.* Southampton: Marine Accident Investigation Branch.

Marine Accident Investigation Branch (2016b). *Report on the Investigation of the Grounding of the Cruise Ship Hamburg in the Sound of Mull, Scotland, 11 May 2015.* Southampton: Marine Accident Investigation Branch.

Norges offentlige utredninger (1991). *'Scandinavian Star' – ulykken, 7. april 1990 HOVED-RAPPORT.* Oslo: Norges offentlige utredninger.

Norges offentlige utredninger (2008). *The Loss of the 'Bourbon Dolphin' on 12 April 2007.* Oslo: Norges offentlige utredninger.

Oltedal, H. A. (2011). Safety culture and safety management within the Norwegian-controlled shipping industry: State of art, interrelationships and influencing factors. PhD Thesis, University of Stavanger.

Perrow, C. (1999). *Normal Accidents: Living with High Risk Technologies.* Princeton, NJ: Princeton University Press.

Pillay, M. (2015). Accident causation, prevention and safety management: A review of the state-of-the-art. *Procedia Manufacturing,* 3, 1838–45.

Qureshi, Z. H. (2007). A review of accident modelling approaches for complex socio-technical systems. *Proceedings of the 12th Australian Workshop on Safety Critical Systems and Software and Safety-related Programmable Systems.* Darlinghurst: Austrilian Computer Society.

Reason, J. (1997). *Managing the Risks of Organizational Accidents.* Aldershot: Ashgate.

Schröder-Hinrichs, J.-U., Praetorius, G., Graziano, A., Kataria, A., & Baldauf, M. (2015). Introducing the concept of resilience into maritime safety. Paper presented at the 6th Symposium on Resilience Engineering, Lisbon, June 22–25.

Slišković, A., & Penezić, Z. (2015). Occupational stressors, risks and health in the seafaring population. *Review of Psychology,* 22(1–2), 29–40.

United Nations (2016). *Review of Maritime Transport 2016.* New York and Geneva: United Nations.

US Chemical Safety and Hazard Investigation Board (2007). *Investigation Report: Refinery Explosion and Fire (15 Killed, 180 Injured).* Retrieved from http://www.csb.gov/assets/1/19/CSBFinalReportBP.pdf

2 The history of safety management

James Parsons and Chad Allen

Introduction

Historically, the maritime world has been lacking a safety culture. It has placed financial gain ahead of crew and environmental safety and has been shaped by reactive measures. Maritime history is documented by numerous examples of safety initiatives that were developed and brought into force after experiencing trends in ship type losses or cargo losses, a significant loss of life or environmental damage (Keefe, 2014). For generations, ship owners and operators have pushed the limits in order to carry more cargo and make more money, with at times a disregard for safe vessel operations. This disregard of safety has been either deliberate, out of greed, or unintentional, simply due to a lack of safety knowledge. Samuel Plimsoll wrote in 1873 that a great number of ships were sent to sea regularly in such rotten and otherwise ill-provided state that they could only reach their destination through fine weather, and a large number were so overloaded that it was nearly impossible for them to reach their destination if the voyage was at all rough. These two causes alone accounted for more than half of shipping losses (Plimsoll, 1873).

The international shipping world is slow to react to the concerns of mariners, and only when there is a disaster typically affecting the public is there enough outcry for the world to take notice and begin to discuss change. Thirty-second news clips of an overturned cruise ship or oil-covered sea birds or other marine life has generally been the spark that has initiated public and political pressure for the necessary discussion to effect regulatory change within the maritime world. International shipping has been regulated through the International Maritime Organization (IMO) since the IMO Convention entered into force in 1958. Although some conventions and treaties were already in place prior to 1958, the IMO has been the mechanism for reviewing, debating and implementing change regarding marine safety internationally since its formation (International Maritime Organization, 2017a). While it can be argued that the IMO has been successful in generating the necessary changes to marine safety through key conventions such as Safety of Life at Sea (SOLAS), International Convention on Standards of Training, Certification and Watchkeeping for Seafarers (STCW) and International Convention for the Prevention of Pollution from Ships (MARPOL), these changes normally only come about as a result of a tragic event. Typically, only upon the receipt of the most recent accident report does the IMO strive to make any necessary changes to the existing conventions.

In addition to the IMO and its three key conventions, in 2006 another United Nations body, the International Labour Organization (ILO), adopted a key convention aimed at improving health, welfare and the quality of life at sea for mariners, the Maritime Labour Convention (MLC). The MLC 2006 condensed and updated 37 Conventions and 31 Recommendations that had already been adopted by the ILO since 1920 (International Labour Organization, 2017).

It is conceivable that the maritime world has reached a turning point and quite possibly we are attempting to prevent the next catastrophe with respect to expected increased shipping in the north with the adoption of the Polar Code. Global warming is reducing the extent of ice in northern waters, making shorter and thus more economical ocean transits much more attractive to potential ship owners and operators. There is a long history of experienced and responsible Arctic operators; however, there is concern that these new options of time and cost saving will entice owners and operators with little or no experience to attempt these uncertain passages. In such a vast and fragile area with insufficient infrastructure and resources in place, the maritime world cannot afford to be wrong with the requirements for shipping as outlined in the Polar Code. Whether the Polar Code marks a turning point of proactivity in the maritime regulatory regime however, only time will tell.

From the beginning

Plutarch writes of Pompey forcing captains to put to sea during a storm, asserting that 'to sail is necessary; to live is not' (Plutarch, 2013). The management of safety in the maritime sector appears to be an afterthought, reactive and forged gradually out of the societal growth, opulence and social well-being provided by the maritime sector itself. Historical accounts of maritime trade date back centuries BC (Casson, 1974; De Souza, 2001; Paine, 2013; Stopford, 2009) and then, like today, a sea adventure was deemed dangerous and rife with peril (Bernstein, 2008). Danger was to ships and their cargoes, with scarce mention ever of danger to the crew, from natural and non-natural perils. No one cared very much for sailors and a ship's crew was only needed to help move the ship and its cargo; thus, the crew were poorly paid and exploited for their physical strength (Paine, 2013). Natural perils include fog, currents, heavy weather, lightning, ice, icebergs, volcanoes, tsunamis, giant waves, sandbanks, rocks, tides and magnetism. Non-natural perils consist of fire, pirates, war, terrorism and strikes (Casson, 1974; Fisher, Jaffe, & Marshall, 2005).

Brooks (1995) in his historical account of the development of lifejackets notes that the life of a mariner was cheap and drowning at sea was merely an occupational hazard, and it took the sinking of the RMS *Titanic* in 1912 to shake up the global maritime community as to the safety value of lifejackets on board ships. Excluding cruise ships, ships are a means to an end, a means used to trade and transport goods and people from one place to another or to execute an operation at sea. Maritime trade and transport were deemed vital for a country's exploration, colonization and growth. Consequently, over the centuries maritime advances were nurtured and developed slowly, with very influential customs and enduring traditions (Boisson, 1999; Paine, 2013; Steele, 1986), and with them practices to prevent loss as well as schemes to limit liability and provide financial protection to ships and their cargoes against damage or loss from the perils of the sea (Chorley & Giles, 1952).

During the 16th century, Northern Europe began to fully realize the value of a strong and robust navy. Navy ships were depended upon for exploration, colonization and the protection of a country and its merchant ships. However, the strength and robustness of a navy were not merely determined by the size and agility of the ships; they were also heavily dependent on the health and well-being of the navy officers and crew as experienced by Royal Navy Commodore George Anson during his torturous four-year mission along the Pacific coast of the Americas in 1741 (Paine, 2013). Subsequent to Anson's 1741–1744 voyage, British and French navies, committed to controlling trade over the vast expanse of ocean from North America to Southeast Asia, realized that in order to effectively operate beyond domestic waters they would have to attend to the health and well-being of the crew, especially with respect to nutrition and recruitment of able seamen (Paine, 2013; Steele, 1986).

The *Oxford English Dictionary* (2005) defines management as the process of dealing with or controlling things or people; it defines safety as the condition of being protected from or unlikely to cause danger, risk or injury. As necessity is the mother of invention, in some respects this attention to the health of navy crews may be deemed an early stage of safety management. In addition to the long-range exploits of Northern European navies during the 16th century, slave and passenger trade at sea experienced exponential growth and consequently the horrors and aversions towards conditions and treatment on board ships garnered increased public attention (De Souza, 2001; Paine, 2013; Steele, 1986). The deplorable physical conditions on board navy ships led to public pressure for improvements; however, the callous behaviour of sailors on board slave, immigrant and coolie ships during the 16th, 17th and 18th centuries did nothing to help their case for better treatment. While the overall living conditions and treatment of crews on board navy ships improved over the centuries, the same was not being experienced on board merchant ships where little or no government oversight was provided, and greed was replacing disease as the biggest threat to passengers and crew (Paine, 2013). During 1870–1872, 1,600 British sailors were jailed for breaking contracts by refusing to sail on what were deemed unseaworthy ships, and 20 per cent of British sailors perished at sea between 1830 and 1900 (Paine, 2013).

While documented practice of marine insurance covering hull and cargo risks dates back to Lombardy, Italy, around 1250, it became popular at Lloyd's coffee house in London in 1688 (Fisher et al., 2005). During the Roman Empire, the reactive action to an imminent fear of a ship foundering during heavy weather was addressed by the jettisoning of cargo overboard, with the loss subsequently apportioned among the owners of the ship and the cargo, a practice said to be taken from the Lex Rhodia (said to be the ancestor of maritime law, circa 800 BC) (Boisson, 1999). Historical accounts from the 12th and 14th centuries indicate that people and slaves could be chosen at random and jettisoned in an attempt to save a ship from foundering through sacrifice (Paine, 2013), a practice similar to that of casting Jonah overboard into the sea to calm its rage (Jonah 1: 15). While the safety of a voyage rested with the captain, sage advice was overridden by ship owners focused more on profit than on the safety of the ship and the crew. To avoid resort to the jettisoning of cargo in order to save a ship suddenly caught in a storm, the Romans proactively adopted the policy of banning sailing during the winter months. The ban was accompanied by an administrative penalty (Boisson, 1999).

Throughout the Middle Ages, rules were implemented for preventing the overloading of vessels by greedy ship owners wanting to earn more from freight, with the first of these load line regulations appearing in Venice in 1255 (Boisson, 1999). Regardless, ships continued to be lost during heavy weather as a result of overloading and recklessness on the part of captains and owners. To further address this, in addition to financial penalties for the careless abandonment of the rules of safe navigation, the Sea Laws of Oleron, the Black Book of the Admiralty and the Consolato del Mar note the decapitation of a ship pilot as penalty for endangering or losing a ship and its cargo (Boisson, 1999; Paine, 2013). The well-established ancient culture of punishment continued throughout the Middle Ages, and rudimentary measures were developed and haphazardly implemented throughout European and Mediterranean regions. However, the era basically ended with various insurance schemes being created and depended upon to protect the financial interest of ship owners and to ensure the safety of maritime trade via ships and their cargoes. Sparse mention was given to the safety of those working on board.

To highlight the influence and power that shipping profits had over safety, it was not until the 1870s, after Samuel Plimsoll's two decades of lobbying and campaigning in the British Parliament against the scandal of 'coffin ships', that measures were provided to the Board of Trade to punish greedy ship owners for overloading their ships by detaining them in port (Boisson, 1999; Paine, 2013; Stopford, 2009). Even then, it took decades for the other major European nations to adopt the British move towards safety at sea. On the western side of the Atlantic, the Jones Act served as federal legislation protecting American workers injured at sea (Jones, 1921). Also referred to as the Merchant Marine Act of 1920, Section 33 of the Act allows qualifying sailors who have been involved in accidents or become sick while performing their duties to recover compensation from their employers.

Given the transient nature of ships sailing within or from one country to another it was relatively easy to evade any punitive regulations or safety measures that may have been established and were in force in various ports or countries. The business of maritime transport continues to present very few barriers to entry and be fiercely competitive. The maritime sector continues to be composed of very varied business mind-sets and operational practices. Profits often tend to be marginal and consequently force owners to be creative in cost-cutting measures, especially when it comes to proactive safety measures. Commercial pressures continue to force seafarers to operate at or beyond the limits of ship safety (IMO, 2013).

Technological advancements during the Industrial Revolution were a significant enabler for societal development and the growth of maritime transport. Maritime networks promoted and helped maintain diverse social, political and economic structures and institutions (De Souza, 2001). The value of international maritime commerce grew exponentially over the centuries, and with it came the need for international cooperation on ways to standardize and control shipping operations. The world continues to be heavily dependent on maritime transport. There is a need to prevent unfair competition gained via ignoring safety regulations aimed at protecting those who work and travel on ships.

The culture of punishment and apportioning blame may be regarded as the genesis of safety culture practice and commenced with exacting payment and enforcing reprisals for damages connected to maritime-related accidents (Kristiansen, 2005). The rationale is that financial punishment would influence human behaviour and consequently result in safer maritime operations (International Chamber of Shipping [ICS], 2013). This approach appeared to have some positive effect in that it resulted in the physical hardware, namely ships, being conceptualized, designed, constructed and maintained to more demanding requirements and thus helped reduce accidents (Thorpe et al., 1997). In hindsight, this approach was a relatively quick and easy fix with discernible results.

While accidents had started to become fewer in number due to vessels being built to higher standards, they were still occurring. These accidents were said to be caused by operational or human error (McCafferty & Baker, 2002; United Kingdom P&I Club, 2003). Subsequently, throughout the 20th century, a culture of safety by compliance to prescriptive regulations began to take shape. This was directed by statutory rules and regulations including the seminal and widely adopted Safety of Life at Sea (SOLAS), Prevention of Pollution from Ships (MARPOL), Standards for Training and Certification of Watchkeepers (STCW) and Collision and Load Line Convention. Such an approach towards safety management was aimed to attack and rectify the weak links and error chains prior to the occurrence of an event. The rationale for compliant safety management was to prevent or mitigate risk before a damaging event occurred (ICS, 2013; Kristiansen, 2005).

Despite the cultures of punishment and compliant safety management, maritime losses remained an unfortunate reality throughout the 20th century as evidenced just by the loss of 167 bulk carriers and 1,352 lives between 1980 and 1997 (Donaldson, 1998). Shipping has continued to be a dangerous industry (Bhattacharya, 2015). Hence, a change in mind-set towards maritime accidents was needed. Consequently, safety management through self-regulation was now the focus, namely self-regulation via the ISM Code but propelled through SOLAS, a mandatory regime (ICS, 2013). ISM takes a goal-based approach to safety and allows operators creativity and freedom to ensure safety is not an afterthought. While allowing for uniqueness in the approach to safety, the ISM Code is nevertheless a reactive approach to safety, as are the earlier approaches. By contrast, the Polar Code, which came into effect on 1 January 2017, appears to be the first proactively initiated, risk-based approach to safety management.

Most recently, during the early stages of the 21st century, the IMO has embarked upon supporting a *just culture* approach to safety management (IMO, 2010). A *just culture* is one built on trust and commitment where seafarers are willing to report incidents, accidents and anomalies without fear of reprisal. However, in order for successful adoption of a *just culture*, the demarcation of acceptable and unacceptable behaviour has to be clearly delineated beforehand (McCafferty & Baker, 2002; United Kingdom P&I Club, 2003). Until such time, we appear to have a *cover-up culture* in which seafarers admit to being afraid of getting caught breaking procedures and consequently cover up their non-conformances or mistakes (Brewer, 2017).

It appears as though we have arrived at a stage in the maritime sector when all past and contemporary approaches to safety management are being combined (Kristiansen, 2005). Figure 2.1 depicts the evolution of the more commonly referred to safety management cultures.

Catastrophic events and their influence

The Titanic *(1912) and SOLAS (1914)*

The RMS *Titanic* was thought to be the largest, safest, most luxurious ocean liner ever built when it was launched in May 1911. The ship, owned by White Star Line, was deemed unsinkable, based on the modern design and enhanced safety features. The *Titanic* met its fate on 14 April 1912 when it sank after hitting an iceberg in the Atlantic, killing over 1,500 people. The follow-up investigation into the sinking found major deficiencies in the construction, safety equipment,

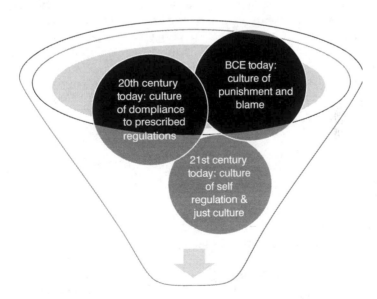

Figure 2.1 Contemporary safety management

communications and crew training. This incident is a defining moment in the history of maritime safety as the International Convention for the Safety of Life at Sea (SOLAS) was established two years later in 1914, and continues to be the leading safety mechanism for maritime transportation globally.

The SOLAS Convention has evolved considerably since 1914 and outlines the minimum specification standards for construction, equipment and safe operations of ships. The Convention governs safety through 14 chapters outlining minimum standards of construction, lifesaving appliances, firefighting apparatus, radio communications and the carriage of cargo, including various hazardous cargoes. SOLAS also focuses on the safe and secure operation of ships and most recently added measures for vessels operating in polar regions as per the Polar Code. Sadly, it took the deaths of 1,517 people for the maritime world to address safety concerns that had long plagued the industry. The SOLAS Convention looked mainly at construction and equipment and little at the human aspect of the *Titanic* tragedy.

Perhaps, also, the *Titanic* was a precursor to what was to come in terms of establishing a connection between the captain of the vessel and the responsible person ashore. During its maiden voyage, the *Titanic* was proceeding at top speed at night and in areas of ice-infested waters, in order to maintain schedule as was the company expectation. Other findings identified during the *Titanic* enquiry targeted shipboard operations which were later addressed by the IMO with the ISM code, although much later in 1998.

The International Maritime Organization (1948)

The effective way to improve maritime safety is through the development of international regulations. Until the establishment of the United Nations (UN) in 1945, there was no efficient vehicle to debate and implement maritime regulations. The Inter-Governmental Maritime Consultative Organization (IMCO) was adopted in 1948 at a convention in Geneva and the Convention entered into force in 1958. The name of the organization was changed from IMCO to the International Maritime Organization (IMO) in 1982. As a specialized agency of the UN, the IMO provided the international approach to the maritime regulatory regime that had global reach. The Mission Statement of the IMO (IMO, 2017b) is to

> Promote safe, secure, environmentally sound, efficient and sustainable shipping through cooperation. This will be accomplished by adopting the highest practicable standards of maritime safety and security, efficiency of navigation and prevention and control of pollution from ships, as well as through consideration of the related legal matters and effective implementation of IMO's instruments with a view to their universal and uniform application. (p.1)

Since the establishment of the IMO, the focus has been to promote safe shipping through the regulations outlined in the various conventions. Even following the *Titanic* disaster and the development of the SOLAS convention, the IMO has a long and documented history of reacting to events rather than attempting to prevent them. Several of the key IMO conventions are a direct result of disasters such as the oil spills of the *Torrey Canyon* in 1967 and the *Amoco Cadiz* in 1978 and the capsizing of the *Herald of Free Enterprise* in 1987, as some of the notable events that led to major changes to the maritime world.

Torrey Canyon *(1967) and MARPOL (1973)*

On 18 March 1967, the Liberian registered oil tanker *Torrey Canyon* ran aground off Land's End in England, spilling over 100,000 tonnes of crude oil into the sea, polluting some 100 miles of British coastline. The accident was caused as the vessel master was anxious not to miss the next tide at Milford Haven, so he attempted a short-cut but misjudged the vessel's position and ran aground at 16 knots, rupturing all six starboard side cargo tanks (Malcolmson, Volk, & Melita, 1967). The regulations at the time favoured the potential polluter with minimum liability for any environmental damage that may be caused. The maximum liability was limited in British courts to $66/tonne of tanker, thus only amounting to $4 million dollars for the case of the *Torrey Canyon* (Burrows, Rowley, & Owen, 1974).

This accident pushed the IMO towards the marine environment and legal issues with the Civil Liability Convention of 1969, activated in 1975. The Civil Liability Convention places the liability for damage on the polluter and also ensures adequate compensation to victims of oil pollution from marine casualties involving oil-carrying ships (Keefe, 2014). The *Torrey Canyon* disaster also led to the International Convention for the Prevention of Pollution from Ships (MARPOL), which is intended to eliminate pollution by oil or other harmful substances and also by minimizing accidental discharges, dumping or exhaust of oil and other substances (IMO, 2017c). The Convention was established in 1973, but was delayed due to the international oil crisis in 1973. The protocol was adopted in 1978 and a combined mechanism entered into force in 1983 (IMO, 2017c). MARPOL consists of six technical annexes covering the Prevention of Pollution by oil, by noxious liquid substances in bulk, by harmful substances carried in packaged form, by sewage from ships, by garbage from ships and also by air pollution.

The *Torrey Canyon* was the worst tanker accident to take place in European waters, resulting in significant economic and political pressures. Again, the IMO was reacting to a disaster that attracted global attention which resulted in new conventions and regulations being applied to the maritime world and focused more on construction and equipment.

Amoco Cadiz *(1978), STCW (1978) and PSC (1982)*

The oil tanker *Amoco Cadiz* experienced a steering gear failure in heavy weather on 16 March 1978 and despite several unsuccessful towing attempts, the vessel ran aground off the British coast, spilling 227,000 tonnes of crude oil, progressively polluting 360km of shoreline. This was considered the largest oil spill in the world at the time, even larger than the *Torrey Canyon* at a relatively similar location 11 years earlier. The spill had a considerable impact on marine life and seabirds, with some species of marine life completely disappearing and 20,000 birds being killed (Centre of Documentation Research and Experimentation on Accidental Water Pollution, 2008).

Public pressure led to considerable updates to MARPOL and SOLAS, including, mandatory towing configurations on tankers, but the *Amoco Cadiz* disaster was mainly responsible for several key changes within the maritime world. The accident prompted coastal states to take a more proactive role in enforcing measures beyond their territorial sea proportionate to the actual or threatened pollution that may result from a marine casualty (Bouyssou, 2016). But more significantly, from a safety management perspective, this accident led to the International Convention on the Standards of Training, Certification and Watchkeeping for Seafarers (STCW) in 1978, which shifted the focus towards the human element of the maritime world.

The STCW convention establishes uniform standards of competence for training, certification and watchkeeping for seafarers on an international level by establishing minimum standards that signatory countries are obligated to meet. The Convention has been amended several times, with the most recent being major amendments in 1995. A unique concept in the 1995 amendments is the proactive nature of the IMO in having countries demonstrate compliance with the Convention, as opposed to simply being a signatory party in the past. Through this new approach, parties are required to present detailed information outlining the measures taken to ensure compliance with the STCW. Each submission is reviewed by a panel of competent persons for approval prior to being added to the list of 'confirmed Parties' (IMO, 2011). This approach by the IMO displayed a deviation from the historical approach of simply changing regulations in reaction to an incident and expecting compliance.

The STCW was amended again in 2010 with continued focus on targeting the human element of the industry through training and certification. These new amendments, commonly known as the Manila amendments, address fatigue, substance abuse, modern technologies, security, education, dynamic positioning, polar waters and changes specific to tanker requirements (IMO, 2011). While fatigue and substance abuse have long plagued the industry, there continues to be a shift at the IMO to address recent changes or trends within the maritime workforce more promptly as opposed to reacting to a disaster.

The wreck of the Liberian-flagged *Amoco Cadiz* also brought another historic turning point in maritime safety with the signing of the Paris Memorandum of Understanding (MOU) in 1982, which led to the implementation of Port State Control (PSC) (Keefe, 2014). Compliance with the various safety regulations fell under the responsibility of the owners, the master and the flag state. As the historical maritime nations tried to tighten safety requirements, owners found an outlet for bypassing these increasingly expensive measures by moving their tonnage to open registries, generally with a more liberal approach to safety. These registries were labelled Flags of Convenience (FOC) and the usage increased continually through the last century as owners were able to save costs on lower taxes, cheaper crews and less verification of safety regulations without any external validation. The implementation of PSC gave port states a device to confirm regulatory compliance of all vessels trading in their waters and provided them the mechanism to detain a vessel that was found in non-compliance. The PSC concept is based on regional agreements to act as a safety net within the region (Chen, 2000). The inspection history of a vessel is transparent to all MOU members and there is an established system of listing vessels according to their inspection results. There are currently nine regional MOUs established to expand the global PSC network. The United States of America, while not a member of any PSC MOU, maintains a similar inspection system to ensure foreign vessels operating in US waters are compliant with international conventions, regulations and treaties.

Herald of Free Enterprise *(1987) and ISM (1989)*

During the 1980s, there was a string of maritime disasters, all building up to another significant change to the safety culture in the maritime world and the shift in focus to the human element. In particular, the sinking of the *Herald of Free Enterprise* minutes after leaving berth in 1987, as a result of mismanagement both onboard and ashore, leaving 193 dead, was the catalyst for the implementation of the International Safety Management Code (ISM) in 1989. The ISM Code is another key advancement in the focus on safety as it provides ship owners and managers with a framework for the development of safety and pollution prevention management systems.

The *Herald of Free Enterprise*, a Ro-Ro passenger and car ferry on the Dover–Calais service capsized after leaving the safe confines of the harbour as the bow door was never closed and no systems were in place to verify the vessel was secured for sea. The accident report concluded that there was gross negligence on the part of both the crew and the owners. The investigation also found that the vessel was overloaded, which was a regular occurrence that the owners were aware of (Ship Disasters, 2016). Lord Justice Sheen in his inquiry into the loss of the *Herald of Free Enterprise* described the management failures as 'the disease of sloppiness' (Department of Transport, 1987). The IMO reacted to this disaster with the ISM Code and the intent to provide those that are responsible for ships with the framework for the proper development, implementation and assessment of safety and pollution prevention management. The Guidelines on Management for the Safe Operation of Ships and for Pollution Prevention were adopted in 1989, the International Management Code for the Safe Operation of Ships and for Pollution Prevention (the ISM Code) and the Code became mandatory in 1998 (IMO, 2017d).

There are several key components of the ISM Code, namely the development of a Safety Management System (SMS) by the entity who has assumed the responsibility for operating the ship. The SMS is a complex and thorough safety system that clearly outlines the master's responsibility and appoints a designated person ashore as the accountable person to ensure the SMS is in compliance. The Code also outlines guidance with respect to resources and personnel, shipboard operations, emergency preparedness, accident and near-accident reporting, planned maintenance systems, auditing process and documentation. To verify that an SMS is in place, the company will be issued a Safety Management

Certificate (SMC) and each compliant company vessel will receive a Document of Compliance (DOC) (IMO, 2014).

The ISM Code is another historical change in the maritime world as the focus continues to be on the critical human factor. The ISM provides all parties with details of accountabilities and responsibilities, detailed maintenance plans for both deck and engine departments, and routine and critical procedures for shipboard operations (IMO, 2014). The requirements and expectations are well documented for all aspects of the vessel operations.

The Designated Person Ashore (DPA) is also a monumental move towards tying the ship-based operations to an identifiable person ashore with access to the highest levels of management. As experienced through the decades of substandard shipping and the Flags of Convenience, vessel ownership and liability can tend to be complex and obscure. The DPA has the responsibility for the company's SMC which is essentially their licence to operate and the responsibility to ensure the fleet under their control is in full compliance of the ISM Code at all times.

Exxon Valdez *(1989) and OPA90 (1990)*

Perhaps the most infamous oil spill in North America resulted from the grounding of the *Exxon Valdez* in March 1989, spilling 11 million gallons of oil into the pristine Alaskan waters, the worst ecological disaster until the Deepwater Horizon blow out in 2010. The cause of the grounding was determined to be a host of errors, namely an ineffective lookout, poor navigational watchkeeping, inoperable radar, reduced manning, lack of shore-side supervision and loss of situation awareness, among other things (Keefe, 2014).

The United States of America (USA) responded by passing the Oil Pollution Act of 1990 (OPA90) which had international repercussions in that it mandated all tankers trading in USA waters to be double hulled. OPA90 also created a comprehensive programme to address prevention, response, liability and compensation for oil pollution incidents within USA waters. It set requirements for crew licensing and manning, created the national Oil Spill Liability Trust Fund, mandated contingency planning and the development of disaster response plans from tankers, tied driving records with mariner licences and mandated post-casualty drug and alcohol tests (Keefe, 2014).

Although OPA90 was an American Act, two following environmental disasters in Europe led to the Eur-OPA, eventually banning the use of single-hulled tankers in Europe. The *Erika* and the *Prestige* both sank and resulted in environmental disasters in 1999 and 2002 respectively. The spills caused outrage and a tremendous political pressure leading to the expedited phasing out of single-hulled tankers (Keefe, 2014).

Why are we not safer?

The earlier IMO conventions, SOLAS and MARPOL, focused on construction and material-based factors, while later conventions and codes focused on the human element with the STCW and ISM. Although the IMO endeavours to make the maritime world safer, we cannot ignore many very recent accidents that in some way escaped the current regulatory regime. While some are related to construction, most are related to human error.

In 2006, the British Columbia Ferries passenger and vehicle ferry, *The Queen of the North*, grounded and sank, leaving two passengers presumed dead. The findings of the accident investigation are generally related to human error which should have been resolved through the STCW and ISM channels; however, the mandatory audits were ineffective in identifying the safety deficiencies (Transportation Safety Board of Canada, 2008).

In 2007, the 4,419 TEU container ship MSC *Napoli* encountered heavy seas, developed a hull fracture and the vessel was abandoned. While under tow, the ship developed a list and was beached for fear of sinking or breaking up. The findings were related to design, heavy weather handling and discrepancies with container weights. Although recommendations were made regarding ship design to the International Association of Classifications Societies (IACS), container ships have been operating for half a century under high structural stresses due the nature of the cargo loading and port rotations. It takes a disaster for these issues to come to light. The maritime world has known about the container weight discrepancies for years; however, the new SOLAS regulations on container weighing only came into effect in 2016 following this disaster (Marine Accident Investigation Branch, 2008).

In 2011, the container ship *Rena* grounded while on approach to the pilot station and eventually broke in two, spilling both oil and containers into the sea. The findings of the accident were related to poor bridge resource management, training and a failure of the SMS, all of which fall under STCW and ISM (Transport Accident Investigation Commission, 2014).

In 2012, the cruise ship, *Costa Concordia* attempted to pass too close to the shore, grounded and eventually foundered, with 32 people dead or missing. The Supreme Court of Cassation upheld a 16-year prison sentence for the master after he had been found guilty of multiple homicide, causing a shipwreck and abandoning his ship while passengers and crew were still on board (Spurrier, 2017). The findings of this accident are targeted at human error in terms of the unsafe navigation and subsequent poor evacuation procedures by the crew, both of which are at the heart of STCW and ISM (Elnabawybahriz & Hassan, 2016).

In 2013, the 8,110 TEU container ship MOL *Comfort* broke in half while transiting the Indian Ocean. Fortunately, all crew safely abandoned the ship prior to the sinking of the forward and stern sections. The ship was only five years old at the time of the sinking and the findings indicate deficiencies with the design and construction for a vessel that size. Inspections of sister ships to the MOL revealed similar hull weakness, and repairs were conducted immediately. The investigation recommends that similar inspections be conducted on vessels of that size or larger (>8,000 TEU) to ensure the structural integrity (Committee on Large Container Ship Safety Japan, 2015). Similar to the MSC *Napoli*, container vessels continue to increase in size and during a typical transit endure considerable stress both in port and at sea.

The *Sewol* passenger ferry capsized and sank in 2014 with 303 dead or missing. During sea passage the ship made a quick course alteration. During the alteration the ship developed a list and eventually capsized. The findings of the accident are related to modifications, overloading of cargo and poor stability management. The findings also cited the poor training of the crew and their inability to handle the emergency. While not directly under the SOLAS convention as it was a domestic vessel, the vessel did come under similar national laws. It is quite clear that no ISM was effectively implemented for this vessel (Kim, 2015).

While some of these disasters are structural or externally related, most of them involve a human element. The IMO has made progress with the introduction of STCW and ISM; however, more work is required on the human element aspect. Human nature is very complex and perhaps the maritime world needs to look at personality types or character traits as a method for determining the optimal personality that can function in a high stress environment as experienced in a maritime career.

Looking ahead to the Polar Code

Following several years of engagement, consultation and debate, the IMO adopted the Polar Code under both SOLAS and MARPOL and it entered into force on 1 January 2017 (IMO, 2017e). While there have been accidents, there has been no environmental disaster or catastrophic loss of life leading to the development of this Code, which really marks a first for the IMO and a considerable turning point in the history of safety in the maritime world. With global warming opening up non-traditional, cost-effective trade routes, particularly the Northwest Passage and the Northern Sea Route, stakeholders initiated discussion on how to avoid a disaster and what was required to try and prevent one in such a fragile environment.

Another component of the Code is the attempt to address both construction and equipment through enhanced requirements and also the human element through manning and training. The purpose of the IMO with this approach is to try and prevent incidents through the proactive mechanism of the Code. Only time will tell if the Code is adequate to prevent an actual disaster, but it is a remarkable achievement for all involved to have completed the process and have it entered into force. If the Polar Code is successful in protecting the polar waters, then this would be a substantial success for the IMO and should affect their approach to all maritime issues in the future.

Conclusion

It has been a long journey from Pompey to the Polar Code, but finally the focus on safety management within the maritime world may be listing to the proactive side. For centuries, shipping was dominated by the wealthy who built, owned and profited from the ships they operated, and financial gain was the number one priority. As the size of financial losses grew and exposure to public scrutiny was increased through substantial spills and considerable loss of life, political pressures mounted for change. Early change for safety focused on vessel construction and equipment improvements, with little regard for crew safety until the creation of the IMO and Conventions such as SOLAS, STCW and MLC 2006. Now more than ever, the human element is at the forefront of safety management. While there is still plenty to improve upon with respect to safety management and leadership, it appears to be steaming towards a more human focus.

References

Bernstein, W. (2008). *A Splendid Exchange: How Trade Shaped the World*. London: Atlantic Monthly Press.

Bhattacharya, Y. (2015). Measuring safety culture on ships using safety climate: A study among Indian officers. *International Journal of e-Navigation and Maritime Economy,* 3, 51–70.

Boisson, P. (1999). *Safety at Sea: Policies, Regulations and International Law*. Paris: Bureau Veritas.

Bouyssou, A. (2016). Maritime disasters trigger regulatory change. https:// greeningmaritimetransport.wordpress.com/2016/04/15/maritime-disasters-trigger-regulatory-changes.

Brewer, S. (2017). Human failure affecting shipping safety performance. http://www. propel.no/news.

Brooks, C. J. (1995). *Designed for Life: Lifejackets through the Ages*. Richmond, BC: Mustang Engineered Technical Apparel Corp.

Burrows, P., Rowley, C., & Owen, D. (1974). *Torrey Canyon*: A case study in accidental pollution. *Scottish Journal of Political Economy,* 21(3), 237–58.

Casson, L. (1974). *Travel in the Ancient World*. Baltimore, MD: Johns Hopkins University Press.

Centre of Documentation Research and Experimentation on Accidental Water Pollution (2008). *Amoco Cadiz*. http://wwz.cedre.fr/en/Our-resources/Spills/Spills/Amoco-Cadiz.

Chen, L. (2000). Legal and practical consequences of not complying with ISM code. *Maritime Policy and Management*, 27(3), 219–30.

Chorley, B. R. S. T. C., & Giles, O. C. (1952). *Shipping Law* (3rd ed.). London: Sir Isaac Pitman & Sons.

Committee on Large Container Ship Safety Japan (2015). *Final Report of Committee on Large Container Ship Safety*. http://www.mlit.go.jp/common/001081297.pdf.

De Souza, P. (2001). *Seafaring and Civilization: Maritime Perspectives on World History.* London: Profile.

Department of Transport (1987). *MV Herald of Free Enterprise: Report of Court No. 8074: Formal Investigation*. London: HMSO.

Donaldson, J. L. (1998). The ISM code: The road to discovery? Inaugural Memorial Lecture to Professor Cadwallader, London, *Lloyd's Maritime and Commercial Law Quarterly,* 1998: 526.

Elnabawybahriz, M. N., & Hassan, M. H. N. (2016). The impact of low efficient evacuation plan during *Costa Concordia* accident. *International Journal of Mechanical Engineering*, 1(5), 43–54.

Fisher, A., Jaffe, A. C. P., & Marshall, M. (2005). *Principles of Marine Insurance*. London: Chartered Insurance Institute.

International Chamber of Shipping (2013). *Implementing an Effective Safety Culture: Basic Advice for Shipping Companies and Seafarers*. London: International Chamber of Shipping.

International Labour Organization (2017). *Maritime Labour Convention, 2006, as amended*. Retrieved from http://www.ilo.org/wcmsp5/groups/public/---ed_norm/---normes/documents/normativeinstrument/wcms_554767.pdf.

International Maritime Organization (2010). *MSC 88/16/1 Role of Human Element: Just Culture-Essential for Safety*. London: IMO. http://www.rina.org.uk/hres/msc%2088_16_1.pdf.

International Maritime Organization (2011). *STCW Including 2010 Manila Amendments: STCW Convention and STCW Code: International Convention on Standards of Training, Certification and Watchkeeping for Seafarers*. London: IMO.

International Maritime Organization (2013). *HTW 1/INF.4 Role of the Human Element: Assessing the Determinants of Safety Culture in Shipping*. London: IMO.

International Maritime Organization (2014). *ISM Code, International Safety Management Code with Guidelines for its Implementation*. London: IMO.

International Maritime Organization (2017a). Adopting a convention, Entry into force, Accession, Amendment, Enforcement, Tacit acceptance procedure. http://www.imo.org/en/About/Conventions/Pages/Home.aspx.

International Maritime Organization. (2017b). Brief history of IMO. http://www.imo.org/en/About/HistoryOfIMO/Pages/Default.aspx.

International Maritime Organization (2017c). *International Convention for the Prevention of Pollution from Ships (MARPOL)*. London: IMO. http://www.imo.org/en/About/Conventions/ListOfConventions/Pages/International-Convention-for-the-Prevention-of-Pollution-from-Ships-%28MARPOL%29.aspx.

International Maritime Organization (2017d). Safety management: Development of the ISM code. http://www.imo.org/en/OurWork/HumanElement/SafetyManagement/Pages/Default.aspx.

International Maritime Organization (2017e). Shipping in polar waters: Adoption of an international code of safety for ships operating in polar waters (Polar Code). http://www.imo.org/en/MediaCentre/HotTopics/polar/Pages/default.aspx.

Jones, W. L. (1921). The Merchant Marine Act of 1920. *Proceedings of the Academy of Political Science in the City of New York*, 9(2), 89–98.

Keefe, P. (2014). Disasters at sea and their impact on shipping regulation. https://www.marinelink.com/news/regulation-disasters371542.

Kim, S. K. (2015). The Sewol ferry disaster in Korea and maritime safety management. *Ocean Development and International Law*, 46(4), 345–58.

Kristiansen, S. (2005). *Maritime Transportation: Safety Management and Risk Analysis*. Oxford: Elsevier Butterworth-Heinemann.

Liberian Board of Investigation (1967). Report on Stranding of *Torrey Canyon*. *International Legal Materials*, 6(3), 480–487.

Marine Accident Investigation Branch (2008). *Report on the Investigation of the Structural Failure of MSC Napoli, English Channel, 18 January 2007*. Southampton: United Kingdom Marine Accident Investigation Branch https://assets.publishing.service.gov.uk/media/547c703ced915d4c0d000087/NapoliReport.pdf.

McCafferty, D. B., & Baker, C. C. (2002). Human error and marine systems: Current trends. Paper presented at the IBC's 2nd Annual Conference on Human Errors, London, 20–21 March.

Paine, L. (2013). *The Sea and Civilization: A Maritime History of the World*. New York: Alfred A. Knopf.

Plimsoll, S. (1873). *Our Seaman: An Appeal*. London: Virtue & Co.

Plutarch (2013). *Complete Works of Plutarch*. Hastings: Delphi Classics.

Ship Disasters (2016). MS *Herald of Free Enterprise*. http://www.ship-disasters.com/passenger-ship-disasters/herald-of-free-enterprise.

Spurrier, A. (2017). *Costa Concordia* captain begins 16-year jail sentence. *Fairplay*, 15 May 2017. https://fairplay.ihs.com.

Steele, I. K. (1986). *The English Atlantic, 1675–1740: An Exploration of Communication and Community.* Oxford: Oxford University Press.

Stopford, M. (2009). *Maritime Economics* (3rd ed.). London: Routledge.

Thorpe, R., McAlear, R., Leback, W., Kendrick, A., Mackey, T., Briggs, R., et al. (1997). The changing role of ship classification societies: Discussion. Authors' closure. *Transactions – Society of Naval Architects and Marine Engineers*, 105, 521–39.

Transport Accident Investigation Commission (2014). *Marine Inquiry 11-204 Container Ship MV Rena Grounding on Astrolabe Reef, 5 October 2011.* Wellington, New Zealand: Transport Accident Investigation Commission. http://www.taic.org.nz/ReportsandSafetyRecs/MarineReports/tabid/87/ctl/Detail/mid/484/InvNumber/2011-204/language/en-US/Default.aspx?SkinSrc=[G]skins%2FtaicMarine%2Fskin_marine.

Transportation Safety Board of Canada (2008). *Marine Investigation Report M06W0052 Striking and Subsequent Sinking, Passenger and Vehicle Ferry Queen of the North, Gil Island, Wright Sound, British Columbia.* http://www.tsb.gc.ca/eng/rapports-reports/marine/2006/m06w0052/m06w0052.pdf.

United Kingdom P&I Club (2003). Just waiting to happen. *Alert!* 2003: 1, 3. http://www.he-alert.org/objects_store/Alert_Issue_1.pdf.

3 Safety management systems

Bjørn-Morten Batalden and Helle A. Oltedal

A historical glance at the ISM code

The era of safety management by the support of safety management systems is relatively new and it is considered to be driven by the findings of the Challenger and Chernobyl accident investigations (Hale, 2003; Hollnagel, 2014). It has entered the domains of the land-based industry, the aviation industry and the maritime industry alike. The objective of safety management systems is to reduce and prevent accidents of managerial origin, by introducing managerial and organizational methods (Okstad & Hokstad, 2001).

Safety management is a continual process of learning at the individual, group and organizational level. Organizational learning is defined as a process based on individual learning that is concerned with creating and providing new knowledge, and adapting it into the organization in a constantly changing environment (Castaneda & Rios, 2007). This involves a dynamic between assimilating new learning and using what has already been learned (Crossan, Lane, & White, 1999).

The demand for learning may be internal or external. Internal demands include a demand for observing and reflecting on everyday work, searching for small deviations and recognizing needs for adjustments, sharing and adjusting stories and accounts of what is going on to build a common understanding and common approaches/reactions to work demands, challenges and unexpected events. Questioning and challenging accepted assumptions and constructs can also be part of the learning agenda (Hollnagel, 2014). External demands include regulations and requirements from authorities, demanding customers and/or suppliers and managing agencies. The subjects of learning appear on different levels, e.g. the technical, operational, social and/or organizational level. For instance, the oil and gas sector is known to have demanding customers with respect to safety, which has led to sector-specific regulations that surpass minimum requirements set by the International Maritime Organization (IMO), such as the Guidelines for Offshore Marine Operations (G-OMO). G-OMO is designed to offer a standard global approach to safe operations. Although these guidelines are external demands for shipping companies, the driving forces were internal demands on behalf of the oil companies. The advantage of such industrial guidelines is that learning in form of changes in regulations occurs far faster than changes in the international IMO regulations. When internal demands are the driving force for changes that support safe operations, it indicates the presence of a safety culture. Safety culture is further elaborated in Chapter 4.

Following the identification of poor management as a contributing cause to several serious maritime accidents in the 1980s and 1990s, the IMO was called upon to develop guidelines concerning shipboard and shore-based management to ensure safe operations of ro-ro passenger ferries (International Maritime Organization [IMO], 2010, p. v). At the time, the International Chamber of Shipping (ICS) and the International Shipping Federation (ISF) had already developed a Code of Good Management Practice in Safe Ship Operation, issued in 1982, which was intended as a guideline only (IMO, 1982). We are often told that the maritime industry implemented such guidelines later than other high-risk sectors, but this shows us that the maritime sector was ahead of its time. However, since the guidelines were applied on a voluntary basis, there were differences within sectors and companies. The ISF/ICS formulation of the overarching principles clearly shows that they were based on contemporary theories of safety management.

> SAFETY and EFFICIENCY are integral to good management. They can only be the result of structured, painstaking policy and a combination of the right skills, knowledge and experience. The direct involvement of decision-taking management in these matters is vital. The attitude of an Owner and/ or senior management is reflected in company policy and thus directly in the work of all the company employees. THE INITIATIVE MUST THEREFORE COME FROM THE TOP.
>
> (IMO, 1982, original capitalization)

These guidelines were later developed to become the International Safety Management (ISM) code for the Safe Operation of Ships and for Pollution Prevention, which was adopted by the IMO in 1993.

The introduction of the ISM code was subject to an intense debate both before and after its adoption, and not everyone was positive about its introduction. Some of the scepticism was directed towards the need for the code itself, its effectiveness and expected beneficial impact (Mejia, 2001). In the years following the implementation, it was quite common to hear opinions such as 'the only practical use of the code is to produce even more paperwork, and its effectiveness is measured by metres of ring binders with procedures'. A survey carried out shortly after implementation shows widespread resistance among seafarers and industry personnel to its obligatory establishment, and it was concluded that the maritime industry was not ready for the ISM code at that time (IMO, 2005).

Now, almost 25 years after its adoption, there is far more research on issues related to the ISM code, its effectiveness, the industry's perceptions of the ISM code itself and its practical implications. A search in Elsevier's abstract and citation database Scopus found 98 peer-reviewed articles published between 1995 and 2017 related to the ISM code. The international maritime regulation became more extensive with the establishment of the International Maritime Organization (IMO) and the International Labour Organization (ILO) (Anderson, 2003). The regulations include the International Convention for the Safety of Life at Sea (SOLAS) of 1974, the Standards of Training, Certification & Watchkeeping Convention (STCW) of 1978 and the United Nations Convention on the Law of the Sea (UNCLOS) of 1982 that codifies flag-state duties.

As mentioned, the origins of the ISM code itself date back to the late 1980s, when there was an increasing concern about poor management standards in shipping, in particular regarding passenger vessels. In the aftermath of accident investigations, major errors on the part of management were identified, and a common factor appearing in these accidents was human error.

Table 3.1 lists passenger vessel disasters happening during the period from 1980 to the *Herald of Free Enterprise* disaster in 1987. For some of the listed incidents (e.g. river boats and sightseeing boats) it was difficult to find ship specifications. Therefore, the ISM code may not actually apply to the entire list. In addition, the number of fatalities are approximate for several of the incidents.

Table 3.1 Passenger vessel accidents precursory to the ISM Code 1980–1987

1980	Bangladeshi ferry *Rushi* capsized by stormy condition at Padma River, 230 people lost their lives
1981	The Indonesian passenger ship *Tampomas II* caught fire and sank in the Java Sea, 580 people died
1981	Brazilian river boat *Sobral Santos II* capsized in the Amazon River, an estimated 300 people died
1981	Brazilian double-decker boat *Novo Amapá* capsized in the Amazon River, about 262 people drowned
1981	A triple-decked river boat capsized and sank in the Amazon River, more than 300 people presumed dead
1983	The Nile river steamer *10th of Ramadan* caught fire after an explosion, 357 people died
1983	Russian cruise ship *Aleksandr Suvorov* crashed into a railroad bridge on the Volga River, 177 people died
1983	*Dona Cassandra* sunk by typhoon Orchid (off Mindanao), 167 lost their lives
1983	Chinese passenger ferry *Red Star 312* capsized, leaving 147 dead
1985	Chinese sightseeing boat sank at Songhua River in China, 171 people lost their lives
1986	The Soviet passenger vessel *SS Admiral Nakhimov* collided with a bulk carrier, 423 people lost their lives
1986	*Atlas Star* sank during a storm on the Sitalakya river, about 500 people are presumed killed
1986	The ferry *Shamia* overturns in the Meghna river during a storm, 600 dead
1987	*Donna Paz* ferry collided with a tanker in the Philippines, an estimated 4,386 people were killed.
1987	*Herald of Free Enterprise* capsized off Zeebrugge, 188 people lost their lives

One of the notable accidents which has been said to lead to the introduction of the ISM code, is the tragic accident of the *Herald of Free Enterprise* in 1987, a ro-ro ferry that capsized only a few minutes after leaving port with the loss of more than 180 lives (Department of Transport, 1987). A formal investigation was launched a few days later by the United Kingdom – the flag state of the vessel – and five months later, the official investigation report was published. The *Herald of Free Enterprise* investigation draws together much that is relevant to human error and organizational management theory in the maritime industry, and sums up the responsibilities for the disaster as follows:

All concerned in management, from the members of the Board of Directors down to the junior superintendents, were guilty of fault in that all must be regarded as sharing responsibility for the failure of management. From top to bottom the body corporate was infected with the disease of sloppiness.

(Department of Transport, 1987, p. 14)

Following the *Herald of Free Enterprise* casualty, the IMO assembly adopted the Resolution A.596 (15) (Anderson, 2003), which tasked the Maritime Safety Committee of IMO with developing guidelines for shipboard and shore-based management. Resolution A.596 (15) recognized that the majority of maritime accidents involved human error and fallibility, which could be traced back to organizational system failures, and that the safety of ships could be greatly enhanced by the establishment of better operational practice. The maritime safety committee was requested to develop, as a matter of urgency, guidelines, wherever relevant, concerning shipboard and shore-based management procedures to better ensure safer operation of passenger ro-ro ferries. Some years later, Resolution A.741(18) which constitutes the International Safety Management (ISM) code was adopted (IMO, 1993) during the 18th session of the IMO Assembly in 1993. This prescribes that shipping companies should ensure that a ship's personnel are able to communicate effectively in the execution of their duties related to the safety management system (SMS) of their ships. The ISM code has undergone some smaller revisions over the years with the most important being in 2010 and 2014. Nevertheless, the code remains relatively similar to the original text.

Safety management and the ISM code today

As a basis for a systems approach to safety management, an SMS typically follows the principles of management systems such as the quality management systems introduced by the ISO 9000 standard with its focus on a process approach (Stolzer, Halford, & Goglia, 2011). An SMS typically contains both reactive and proactive elements. The proactive elements are risk assessment while the reactive elements relate to reporting and learning. As such, the SMS facilitates the objective of safety management as the systematic actions taken to ensure that activities follow policies, objectives, and other requirements set by a company. Safety management is described by Aven and Vinnem (2007) as a collection of all activities 'designed to direct and control an organisation with regard to safety'. The objective of safety management is to regulate or control activities where management direct processes to ensure safe conduct (Hollnagel, 2014). This objective typically constrains the operational freedom of operators and as such tries to limit uncertainty. Of course, an SMS can also be developed to facilitate the capacity to manage uncertainty. The ISM code leaves the choice as to how the SMS shall be developed to the companies, but there has possibly been an assumption among companies that instructions, procedures and checklists are to be steering documents. Today the shipping industry is seen as a highly proceduralized sector.

According to the ISM Code, a safety management system should contain the capacities for assisting in the management of safety. The following mechanisms should be in a safety management system for ships based on the ISM Code (IMO, 2010):

- a safety and environmental protection policy;
- instructions and procedures to ensure safe operation of ships and protection of the environment in compliance with relevant international and flag state legislations;
- defined levels of authority and lines of communication between, and amongst, shore and shipboard personnel;
- procedures for reporting accidents and non-conformities;
- procedures to prepare for and respond to emergency situations; and
- procedures for internal audits and management reviews.

Figure 3.1 outlines the main characteristics of an SMS as required by the ISM Code. An SMS is a 'structured and documented system enabling Company personnel to implement effectively the Company safety and environmental protection policy' (IMO, 2010, p. 10). The circle symbolizes the continuous learning which is a central element in safety management. At the core of the SMS are the policies that state company values and norms. Hence, the policies should provide guidance for the content and structure of the SMS. A main driver in the development of the SMS are rules and regulations together with codes, guidelines, standards and best practices. In this model, these are gathered under the term regulations. Another important element of an SMS based on the ISM code is the ability to gather information from the operations, which is called operational experience information in the model. The information from operational experience is analysed to provide input to decision processes. Based on the analysis, decisions are made with the objective of ensuring an appropriate level of safety that should be in line with the overall policies.

In the following sections, the elements depicted in Figure 3.1 will be further elaborated, starting with the shipping environments and regulations.

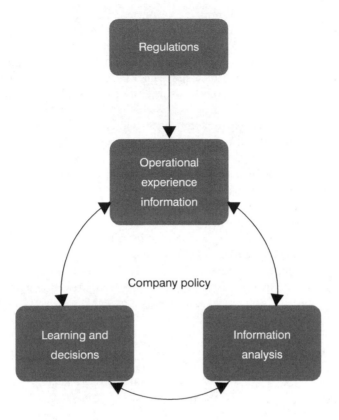

Figure 3.1 The core elements of the safety management system according to the ISM Code

The shipping environments and regulations

An SMS based on the ISM code is obliged to ensure that ship operations comply with mandatory rules and regulations. Further, the SMS should take into consideration different codes, guidelines, and standards within the maritime industry organizations (IMO, 2014, section 1.2.3.2). These can originate from IMO, flag states, classification societies and other maritime industry organizations such as the Nautical Institute, the International Chamber of Shipping and the Marine Safety Forum. At the same time, the companies are given a certain freedom in how they comply which is the goal of an enforced self-regulatory regime such as the ISM code. The shipping industry is a heterogeneous industry, and this is to assist companies in finding the approach best fit to their own operations.

Baldwin, Cave and Lodge (2012, p. 15) state that the need to regulate follows mainly because of 'market failure' or even 'market absence'. 'Market' refers to a failure of production expected by society; for example, a failure to establish a safe and healthy workplace or the safe production of maritime transportation. The ISM code is founded on a special form of regulation, enforced self-regulation, which 'involves a subcontracting of regulatory functions to regulated firms' (Baldwin et al., 2012, p. 146). In contrast to a pure self-regulatory regime where a company or organization is allowed to make their own internal regulations without interference from society, enforced self-regulation allows for the same but with certain limitations and control mechanisms enforced by society. Two of the main advantages of self-regulation relate to how the regulatory mechanism allows for expertise and efficiency. Expertise relates to the relevance and proximity of the industry that it regulates. It is beneficial for the ones writing the regulations for an activity to have in-depth knowledge of the activity to be regulated and first-hand experience of the activity. A person with limited knowledge and who has little experience might write regulations that are not appropriate for the activity. If the regulations are not seen as appropriate for the operations at hand, the people involved may lose trust in the overall SMS; it is possible that the company would enter a vicious circle with increasing non-conformities, violations, underreporting of experience data and the like.

If the regulations of activities are developed by experts, a heightened level of compliance can be expected. This effect is supported by research in the maritime industry (Antonsen, 2009; Ripamonti & Scaratti, 2015). It also provides better knowledge on the need for regulation and as such reduces the costs of preparing rules, instructions, procedures, etc. An enforced self-regulatory regime can reduce the enforcement cost on society as the companies are obliged to control their activities – shifting the cost to the users rather than to taxpayers in general. The efficiency of controls may also increase, as the experts who have written the regulations know what to look for and when to look. On the other side, there are negative issues with an enforced self-regulatory regime. The term capture is used to illustrate the risk of biased behaviour in managing enforced self-regulation. It is possible for companies to regulate their activities to boost economic results rather than for safety purposes. Thus, they would fail in their duty to protect the public interest.

Four characteristics/dimensions present themselves in regulations providing an analytical dimension to understanding ways of regulating:

1   target – the entity that a regulation is relevant to;
2   regulator – who creates and/or enforces the regulation(s);
3   command – whether a means (mandate or prohibition) or end (target to achieve or avoid) function is being applied; and
4   consequences – being either positive or negative.

Enforced self-regulation relates to the 'unity of regulator and target' and the government's requirements to regulate activities which it will somehow monitor (Baldwin, Cave, & Lodge, 2010, p. 150).

The ISM code recognizes this in its preamble (IMO, 2010) by stating that no entities operating ships are the same and thus need a degree of freedom in developing the SMS. While, for example, STCW sets forward requirements on what is to be done and some general requirements on how it shall be done, the ISM code leaves the companies a certain liberty on how the requirements are to be complied with. STCW requires the master to determine the watchkeeping arrangements, taking into 'account all pertinent circumstances and conditions' (IMO, 2011, p. 353). Still, it says little that is more specific, and this is one example where the company can (and should) develop instructions and procedures that are more precise in their own SMS on how this should be done to enhance the safety of navigation.

In some parts of the maritime industry, there are no international or national rules and regulations to address what and how operations are to be conducted. One example is operations using dynamic positioning where there are no international requirements. The maritime industry regulated this part of maritime operations for decades through the non-governmental organization the Nautical Institute (Bray, 2008). This highlights the positive role self-regulation can have in safety management. The industry had incentives to self-regulate this operation because it involved critical operations where a serious accident would probably result in regulatory measures that might have limited the industry's opportunity to self-regulate. A risk of being heavily regulated is that it might limit, for example, innovation.

The maritime industry is heterogeneous in the sense that ship design varies between different segments, and the same goes for many operations. Ships are built for very different purposes and thus sometimes need different ways of operating. Some ships are very specialized and certain ships fall outside the existing rules and regulations. An example is what has become special purpose ships (SPS). The IMO made the SPS code to deal with ships that fall outside the definitions of the SOLAS convention. The international and national regulations develop slowly through the working groups of the IMO and in the international community, and have limited capacity to 'cover' new solutions and innovation. At the same time, the shipping industry often sees opportunities in new markets. Only after the anchor handler tug and supply vessel *Bourbon Dolphin* capsized and sank in April 2007 were new regulatory requirements developed with respect to issues such as the onboard stability calculator. The ISM code allows the maritime industry to strive for managing operations that are not covered by existing rules and regulations, but it also provides them with the opportunity to self-regulate with objectives other than ensuring safety.

The degree of external threats plays a significant role. The external threats are divided into economic, social and regulatory, providing incentives to comply. A self-regulatory regime will only function when the targets understand and decide that it is in their own best interest to comply. In the maritime industry, it is expensive to be held back in port due to non-compliance with the requirements of the ISM code. This may occur either if the flag state or recognized organization withdraws the safety management certificate or identifies major non-conformities that have to be rectified before leaving port, or if the port state control finds reasons to place the vessel under detention. It is, however, difficult for these organizations to find objective evidence that ship operations are not in compliance with the ISM code. This can be due to lack of expertise in a specific type of vessel or the time available for inspections.

As mentioned, the ISM code falls under the definition of enforced self-regulation meaning that there is an external body that assesses whether the company is self-regulating their activities. Through the ISM code, this task of enforcement is allocated to the flag state or a recognized organization approved by the flag state. The recognized organization is typically a classification society such as DNV GL, Lloyd's Register or the American Bureau of Shipping (ABS). A main challenge in the maritime industry is that it is heterogeneous, thus making it difficult for anyone without in-depth knowledge of the different segments and adhering operations to scrutinize the SMS. This provides an opportunity for companies, whose objective may not be enhanced safety but rather economic performance, to regulate for increased profit. Only after an accident will the companies with this agenda come under scrutiny.

Company policies

Each SMS has one or several policies (presented by the company) as a core to provide foundation and guidance to the content and structure of the SMS. Typically, there is a health and safety policy, an environmental policy and a quality policy. Sometimes companies combine these policies.

With a safety management system, an organization typically develops one or more safety policies to present the organization's safety objectives and its strategy for achieving these objectives – it provides the organization with 'global' criteria. The ISM code requires that the policies cover the objectives of the code stated in section 1.2 (IMO, 2010). By providing the organization with global criteria for judging the level of safety, it should enable the organization to identify 'specific standards which govern particular areas of operational practice' (McDonald, Corrigan, Daly, & Cromie, 2000). For example, in the maritime industry, such standards might be the standard for conduct of navigation or cargo operations. These standards may be described in instructions and ship-specific standards.

The policies are perhaps the main documents to link the SMS to the safety culture of the companies. The concept of safety culture is further elaborated in Chapter 4. As stated in the preamble of the ISM code, commitment from the top management is the 'cornerstone of good safety management' (IMO, 2010, p. 10). The policies can be seen as a stabilizing factor in an otherwise challenging and volatile commercial environment. It is where the top management sets the core values and norms, and establishes the basic assumptions of the company. Typically, this is expressed by a commitment to ensure and maintain the health and safety of all employees, followed by some objectives such as zero accidents, zero personal injuries and zero damage to equipment, etc. The policies should ensure that the norms, values and basic assumptions are not influenced by changes in the economic market. The role of the SMS is to make these norms and values an actual part of the company, and that the conduct of business is in line with the basic assumptions of the company.

In their study of container terminal operators, Lu and Yang (2010) found that safety policies had a positive influence. Hadjimanolis and Boustras (2013) found that safety policies have a direct impact on both safety performance and the safety climate. Further, they claim that safety policies also contribute to 'increase job satisfaction and organizational commitment' (Hadjimanolis & Boustras, 2013, p. 55). Findings from Batalden and Sydnes (2014a), however, indicate that few maritime accidents can be directly connected to the policies. Only 0.4 per cent of the 478 causal factors in the study relate to the safety and environmental protection policies (Batalden & Sydnes, 2014a). This study has clear limitations as the accident reports used were coded without any attempts to assess the level of details in the accident investigations. A more thorough study of the policies of the companies involved might find that there are weaknesses in the policies, or that there are no clear links between policies and the other SMS activities.

Operational experience information

A cornerstone of safety management, and therefore to the capacity of an SMS, is the ability to learn from incidents and accidents. The ISM Code relies on a linear causality with the assumption that future accidents can be predicted and avoided by adjusting the organization and operations based on input from previous operational experience. This requires companies to measure their activities. Data input might be quantitative measures, e.g. how equipment is used such as engine control, or other accumulated data from operations, or it can be input from qualitative measures such as reports, or more open forum discussions such as officers' conferences; some companies also collect suggestions for improvements.

The main mechanism used in the maritime industry is crew member reporting of non-conformities to rules, regulations and the SMS, as well as hazardous situations and incidents (IMO, 2014, sections 9 and 10). The reports are to be investigated and analysed, providing corrective and preventive actions for the future. Kjellén (2000) suggests that if the data input from these reports are reliable and accurate, it is possible to adjust or change the SMS and thus control operational safety. Oltedal (2011) identified substantial challenges in the aspects of both reporting and collecting necessary data input, meaning unreliable data processing and analysis, poor development of safety measures and lacking implementation of the safety measures. Reasons for this underreporting are a fear of negative impact on crew members, a system that thrives on compliance, complicated reporting systems that make it difficult for crew members to report and a lack of overall SMS understanding. The discrepancy between what is written in instructions and procedures of the SMS and what needs to be done or what is believed needs to be done is highlighted as a main reason for poor reporting (Oltedal, 2010). In their study of incident investigation systems, Jooma, Hutchings, Hoagland, and Jandrell (2016) found that investigations focus on local levels – between personnel within the same department. They suggest that the investigations should have a broader perspective, including the roles other parts of the organization and outside the organization might have on the incident.

Another challenge is the lack of relationship between minor incidents and major accidents as mentioned in Chapter 1. There is not necessarily a link between these two. Still, many SMS focus on the reporting of as much as possible with the objective of being able to adjust the SMS and operations to avoid these issues in the future. This may lead to overload at the department that analyses the reports. Too many reports on minor issues dealing with personal safety, such as use of helmets, endanger the safety management related to process safety. See Chapter 1 for an explanation of the difference between personal safety and process safety.

In their study of the manufacturing industry, Granerud and Rocha (2011) find that it is possible to facilitate learning in organizations, but that advanced continuous learning in an organization is not facilitated within occupational health and safety management systems. For more advanced learning, there is a need for managerial support in the organizations that facilitate employee involvement, receiving suggestions, and experimentation and discussions (Granerud & Rocha, 2011). A similar situation might be present in the shipping industry. If a company operating ships limits the involvement of crew members in the development of internal regulations such as instruction, procedures and checklists, the company limits their continuous learning to information from reported incidents and accidents. The more tacit knowledge of operations is perhaps not taken into account and valuable success criteria are not part of the input to these internal regulations. If companies fail to facilitate this, internal regulations may develop that describe work as imagined rather than work as done (Hollnagel, 2009), which increases the likelihood of deviation from these internal regulations.

Information analysis

Information analysis is central to a functioning SMS. The objective of information analysis is to facilitate the making of sound decisions within a company. The ISM code requires companies to analyse information and use the results from the analysis to improve safety. In this process, the companies are to ensure that both corrective and preventive actions are taken in a timely manner. The ISM code emphasizes the use of risk assessment and the analysis of risks. This corresponds to the pillar of risk management typically found in safety management system approaches. The objective of risk analysis is to map and describe risks; the risk analysis provides a risk picture for the decision makers and users.

According to the code, all companies must establish safeguards for all identified risks and as such set out the requirement for risk assessment. This indicates that companies should identify all risks, which is probably an impossible task, taking into consideration problems such as the unknown unknowns (Aven & Zio, 2014). It should, however, be read to mean that companies should continuously strive to identify all risks, and that safeguards should be established for identified risks. In the 2010 edition of the ISM code the requirement for risk assessment was made more explicit so that risk assessment became an important part of safety management (IMO, 2010). The risk assessments are the input to the content of the

SMS – an example is the development of instructions, procedures and checklists for both the shore administration and each vessel in a company.

In risk management, the dominant approach has been to measure risk, while the issue of uncertainty has been less in focus. Recently, uncertainty has become important in addressing rare events that are difficult to detect (Grote, 2015). Research into how one may address these uncertainties calls for new methods and new approaches. Grote (2015) in her paper 'Promoting safety by increasing uncertainty: Implications for risk management' discusses the role of handling uncertainties in risk management and argues that it sometimes might be beneficial to maintain or even increase the uncertainty when conducting risk management to improve safety. By 'making deliberate operational and strategic choices between reducing, maintaining, and increasing uncertainty in order to establish a balance between stability and flexibility in high-risk systems while also matching control and accountability for actors involved', more effective risk management can be conducted (Grote, 2015, p. 78).

Where this type of system is practised (Grote, 2015) is:

1   *High reliability organizations*, such as aircraft carriers and nuclear power plants where maintaining uncertainty is accepted by the organization, as uncertainty is managed collectively by the decentralization of decision-making.

2   *Ultra-resilient systems*, which seek uncertainty as part of their business model. The safety of such systems heavily rely on the expertise and competence of the actors. Combat operations is an example of where one can find these systems.

It is argued that the prime focus of classical risk management is to reduce uncertainty so that one reaches an acceptable risk level based on some criteria. In such instances, instructions, procedures and checklists will enhance standardization and reduce uncertainty. This may result in similar, repetitive and restricted behaviour, with limited opportunities to experiment and find innovative and possibly safer solutions. Risk perception is discussed in Chapter 6.

Studying high reliability organizations, decentralization of decision-making by empowering local agents is central and thus does not necessarily seek standardization with a very strict decision envelope. In such organizations, flexibility is key and rather than reducing uncertainty, they are willing to maintain a level of uncertainty to gain this flexibility. Uncertainty may have to increase to maintain flexibility not only when handling disturbances to a system but also to facilitate innovation. By allowing agents to self-organize, it may result in new ways of conducting operations instead of limiting them to existing best practices. This is not to say that risk assessments should always be left out prior to the development of new practices, but that the flexibility to improvise and self-organize decisions and operations will increase uncertainty at least for a period of time.

In managing risks in organizations, these options need to be understood and accepted in order for the organizations to choose the most suitable. Further, two fundamental criteria need to be taken into consideration (Grote, 2015):

- optimizing the balance between stability and flexibility, and
- optimizing between control and accountability.

One important task for an organization is to balance the mix between stability and flexibility where 'flexibility is a response to uncertainty, while stability is a response to the need for control' (Grote, 2015, p. 73). In optimizing control and accountability, the objective is to only hold actors accountable for those outcomes they actually can control. This implies that actors should be held accountable for outcomes that they actually can control and that each actor should be responsible for those outcomes that lie within their role to control. As such, all levels from regulators, managers, supervisors, designers, etc. to the operator at the sharp end are accountable for their actions. Furthermore, the management of uncertainty needs to be addressed in the context of power relations, which can transfer uncertainty to less powerful agents (Grote, 2015).

Studying the offshore support vessel segment, Batalden and Sydnes (2014b) addressed how and to what extent companies applied risk analysis as part of regulating their activities when developing their SMS. The companies rely heavily on external inputs from rules and regulations as well as charterers and industry best practice. Risk assessments are not applied systematically to develop instructions, procedures and checklists. A longitudinal study conducted by Borch and Batalden (2014) found that only a few risk assessments were conducted prior to a fleet of offshore support vessels entering into a new challenging operation area with new operational elements.

Decisions

When operational experience data is gathered and analysed, the compiled information is distributed to decisions makers in the organization to support the development of remedial actions. The company is free to decide what kind of action they find most fit and, as earlier mentioned here and in other chapters as well, the shipping industry is known to be a highly proceduralized sector, not only extending the use of procedures, but the accompanying checklists as well. Oltedal (2011) found that the industry emphasizes the development of standardized safety measures in the form of procedures and checklists. Organizational root causes related to company policies (e.g. crewing policy) are identified and addressed to a lesser degree.

Hale and Borys (2013) suggest two approaches in the development of procedures – the top-down, rational approach and the bottom-up, constructivist approach. With the top-down approach, procedures are considered more static and violation of these procedures is seen to threaten safety. These procedures put clear constraints on the operators. Procedures based on a bottom-up approach are more dynamic and local, allocating more power to the expert, allowing more variation to be taken into account.

A challenge is the difference between work-as-imagined and work-as-done (Hollnagel, 2009). Often people find that procedures constrain their way of working. Typically, employees adapt their own ways of doing things in the absence of procedures and the same goes for procedures that do not reflect the actual tasks. For these gaps to diminish, processes need to address the institutionalization of the procedures. It is important to ensure that there is good communication with the personnel at the sharp end of operations so that reasons for procedures are understood (interpretation), and they are accepted and effective (Benn et al., 2009). It is also important to facilitate feedback and possible adjustments or changes to the procedures. Some companies within the aviation industry make great efforts to achieve institutionalization through training and simulator exercises. Batalden and Sydnes (2014a) found that many companies in the maritime industry believe these changes will happen with limited or no training.

In January 2015, the pure car and truck carrier (PCTC) *Hoegh Osaka* developed a severe list and was left stranded (MAIB, 2016). The incident report reveals that rules were violated and procedures not followed. However, the report also addressed the drawbacks of uncritically mitigating risk with procedures and checklists and the like. This is also addressed in Chapter 1, with reference to James Reason (1997). *Hoegh Osaka* had a significant number of procedures and the accident can at least partly be blamed on the crew finding them difficult to use. An example is the lengthy checklist that was applicable for cargo operations with 213 tick boxes in total (Marine Accident Investigation Branch, 2016, p. 25). All the boxes were ticked by the chief officer but following the accident investigation it was concluded that the extent of the checklists made it difficult for the crew to identify the more important issues for the safety of personnel and protection of the environment. It seems that the company had a top-down approach, providing the crew with little freedom to decide how things should be done based on their expertise. Rather, it is likely that the company expected that safety was ensured by listing 'everything' that had to be remembered and that the crew was to comply with this. This is different from proceduralizing in the aviation industry. For a Boeing 737 there is an extended checklist used during training and a shorter checklist that is used in operation. If in doubt, the pilot can use the long checklist in operation. The short checklist has been developed by selecting only the most important items to keep the operation safe. Typically for the aviation industry, there is a philosophy for developing checklists and NASA has conducted several research projects to arrive at this philosophy (Degani & Wiener, 1990).

In a study of the offshore support vessel segment, Batalden and Sydnes (2014b) find that procedures are developed based on external input from rules, regulation, guidelines and directions given by the charterers. Few companies use a risk-based approach when designing their procedures. In her study, Oltedal (2011) finds that procedures and checklists are often perceived as being problematic to use in daily shipboard operations, and that they do not reflect the operational situation. Oltedal (2011) suggests that the seafarers' experience be taken more seriously, with regard to both the reasons for underreporting and their experience with new safety measures.

Safety management and organizational learning

In order to be able to carry out efficient safety management, both the individuals and the organization need to learn. Organizational learning is key to knowledge creation in an organization (Yang, Fang, & Lin, 2010). In their study on the transfer from experience to knowledge, Argote and Miron-Spektor (2011) found that most definitions of organizational learning relate to the changes in an organization following acquisition of knowledge and that this change is a function of experience. Further, 'organizational learning occurs over time' (Argote & Miron-Spektor, 2011, p. 1124). Huber (1991) identified four constructs for organizational learning – which also are important pillars for safety management.

- acquisition of knowledge
- information distribution
- information interpretation
- organizational memory.

Five sub-constructs are present within acquisition of knowledge – (1) congenital learning, (2) experiential learning, (3) vicarious learning, (4) grafting and (5) searching or noticing (Huber, 1991). Few, if no. organizations start without some kind of pre-knowledge representing the congenital learning. The members of an organization carry with them certain knowledge when entering the organization. A new shipping company will need to employ already certified and experienced crew members on board their ships. These new crew members will bring previous knowledge acquired from their former employment. This may be both a hindrance and opportunity when establishing procedures. The experience held by new crew members can contribute to the development of the SMS if the company manages to extract this knowledge. At the same time, the pre-knowledge can hinder changes in conduct.

Once an operation is running, companies will gain knowledge through experiential learning, this being either through systematic work or through more unintentional work (Huber, 1991). The role of an SMS is to enhance the capacities for gathering running experience in a systematic way. Organizational experiments are one way of achieving experiential learning. This requires the companies to admit that there is some uncertainty while they at the same time have to show determination (Huber, 1991). One way of establishing organizational experiments is to present a solution to an identified problem, and ensure feedback from those that are influenced by the change (Huber, 1991). An SMS needs to have the capacity to allow the members of the organization to provide feedback to the SMS both from negative and positive experiences.

Another way of acquiring knowledge is through other organizations' learning (vicarious learning). Borrowing from other organizations may be efficient with respect to both cost and enhancing safety. Aligning themselves with other companies reduces the likelihood of sanctions by stakeholders. Grafting is the acquisition of knowledge through hiring new personnel that can bring with them knowledge, skills and expertise that are presently not held in the company. In the search for knowledge, companies can achieve this either by scanning the industry, carrying out a more focused search within a specific area, or through performance monitoring. When scanning, companies search broadly within the industry and segment to identify important knowledge that has not been experienced in-house. This can be done through participation in conferences or monitoring the media. A more focused search is when members of the company study narrower issues such as design input to vessels either internally in their own organization or within their segment. It may also be a search on operational practices. Based on their own policies and objectives, companies can monitor their performance. This is typically done in shipping through internal and external audits. In addition, shipping companies often use key performance indicators to monitor their achievements compared to their own objectives.

For organizations to learn, it is important that the knowledge accumulated in the company is distributed to relevant stakeholders. This sharing of knowledge requires suitable methods for distribution, which can be a challenge in the maritime industry due to limited communication with the vessels. For example, with non-standardized ship designs, it may be difficult to select what to distribute to whom. It will possibly require resources and there will often be the dilemma between sharing important information and at the same time ensuring that the organization is not overloaded with information. New software solutions for SMS may reduce this challenge. Many of these systems help the company to make predefined receivers.

When information is distributed, the interpretation is important. Huber (1991) raises the issue of whether it is best for the organization to interpret the information similarly or if it is better for the organization's different units to interpret the information differently. An SMS should be, somehow, capable of detecting the interpretation. If the company communicates interpretations that vary, it may further enhance organizational learning.

Organizations have a limited capacity to memorize information. Elements that influence organizational memory are (Huber, 1991):

- high personnel turnover;
- the uncertainty related to what information is needed in the organization;
- that personnel in need of information do not know where to retrieve the information.

The SMS needs to address these issues in order to function well. For instance, crew turnover is a challenge for many companies due to the crewing strategy chosen. Often, companies sign on crew members for a certain period, and they may never return to the vessel or even the company. Newer software solutions for SMS have intuitive search capabilities that will reduce the challenge of finding relevant procedures. Still, not all vessels have these solutions. With the typical paper-based systems, this challenge grows with the increasing size and scope of the SMS.

Learning arenas are physical or virtual places where social-cognitive learning processes take place 'here and now', or are constructed and/or supported (Nonaka, Toyama, & Konno, 2000). They can be formal or informal and occur within and between organizations. They demand interaction and activity, and the participants are not necessarily permanent. Physical organization is important. Formal learning arenas could be supervision, guiding, coaching, team/project organizing, job rotation, trainee programmes, observation, networks, training courses, simulator training and meetings. Informal arenas could be lunch chats, corridor chats, asking colleagues for advice, walk and talk, informal networks and self-organizing teams/groups. Learning arenas can be established in relation to operations and upcoming events and changes. The time span is from strategic preparation meetings a long time in advance to safe job analysis (SJA), safety meetings and toolbox meetings just prior to operation. The optimal learning organization cultivates formal and informal learning arenas whenever possible and appropriate.

Opportunities and hindrances for learning, as well as learning both the right and the wrong way, could be present on different levels in the organization (Von Krogh, 1998). The challenge for organizations is to initiate, motivate for, support and ensure that constructive learning processes are always ongoing at all levels in the organization, even beyond the boundaries of organizations: boundless learning (Tharaldsen et al., 2013).

Does the ISM code facilitate better safety management and the way forward?

There are studies that indicate that safety, safety culture and safety management has developed in the right direction after the introduction of the ISM Code (IMO, 2005; Kongsvik, Størkersen, & Antonsen, 2014). However, none of them conclusively shows that the improvements are the result of the ISM code (Heijari & Tapanainen, 2010). Some key issues that are needed to enhance safety at sea are presented in this section. Following the discussion earlier in this chapter, the prevailing form of safety management systems can be considered somewhat limiting. This is not to say that all safety management systems have major weaknesses.

Studies have indicated that people are more willing to learn from accidents rather than from near-misses, and that risk assessment is used to a limited extent. If this is the case, safety management systems as we know them have only a limited capacity to manage safety. If the maritime transportation industry is to incorporate flexibility into their organizations, as Grote (2015) claims is beneficial to enhance innovation, this requires thorough and extensive training of personnel. In many segments of the shipping industry this is not the case, as identified by Batalden and Sydnes (2014a), Bhattacharya (2012) and Oltedal (2011). In such cases, safety management systems need to control the activity until the necessary level of expertise is achieved. The safety management system needs to address the issue of expertise, providing management with information to take the necessary actions to achieve a high level of safety – safety being the outcome of good performance, not only from experiencing near-misses and accidents – see Chapter 1 and the Safety-I and Safety-II discussion. The SMS needs to inform management of the status, enabling them to change the system so that it can withstand and handle emerging phenomena that are difficult to foresee. This may be achieved by increasing the level of flexibility. Companies are confronted with a duality issue where the management must facilitate training and experimentation in learning while at the same time ensuring a certain level of conformity.

The heterogeneous characteristic of the shipping industry not only allows for capture, it is also a limiting factor when companies seek advice in the development of the SMS. Unless the flag state or recognized organization allocates personnel with competence in both safety management and specific knowledge of the operation(s) relevant to the companies, there are limited opportunities for external input and assessment of the SMS (Batalden & Sydnes, 2015). Within the offshore support vessel segment, the external audits carried out by classification societies focus mainly on compliance with existing rules and regulations and technical elements of the ISM code. There is little addressing the relevance of established safety measures in the SMS. This situation might be a hindrance to the development of the SMS.

So, is there any possibility of enhancing safety by the use of an SMS? The ISM code falls well within the safety perspective of Safety-I (Hollnagel, 2014), where the companies establish an SMS that monitors the activity, measures deviations and eliminates these to achieve safety. There are dilemmas with this approach. The model under the Safety-I perspective is linear, assuming that there is a clear connection between action and consequence. This model uses incidents and accidents as a source and as measurement criteria to adjust the SMS, but ignores (at least to some extent) all the variations that result in successful and potentially safer operations. As companies experience fewer incidents and accidents, there is less available information to manage safety. Methods such as hierarchical task analysis (HTA) are developed that can assist the development of e.g. soft barriers such as instruction, procedures and checklists. It seems that such methods are not used in the shipping industry. The Safety-II perspective assumes that there are many different successful ways of conducting an operation. These different alternatives are needed in order to have a resilient and robust system that is able to absorb variations in operations. Within resilience engineering, a method exists to aid the development of an SMS enabling the companies to address the issue of variability. This framework is the Functional Resonance Analysis Method (FRAM) (Hollnagel, 2012). FRAM is a method to address the dynamic and sometimes complex socio-technical systems. It enables the users to understand why incidents and accidents sometimes occur while also being able to explain why operations are successful. Central to FRAM is the continuous effort to adjust performance so that it corresponds with the conditions.

For future research, we encourage the maritime industry to explore the Safety-II perspective, with less emphasis on control strategies in the form of standardized procedures and checklists, supporting variability and flexibility and the development of skills and competences needed to make decisions under uncertainty.

References

Anderson, P. (2003). *Cracking the Code: The Relevance of the ISM Code and its Impact on Shipping Practices*. London: Nautical Institute.

Antonsen, S. (2009). The relationship between culture and safety on offshore supply vessels. *Safety Science*, 47(8), 1118–28.

Argote, L., & Miron-Spektor, E. (2011). Organizational learning: From experience to knowledge. *Organization Science*, 22(5), 1123–37.

Aven, T., & Vinnem, J. E. (2007). *Risk Management with Applications from the Offshore Petroleum Industry*. Dordrecht: Springer Verlag.

Aven, T., & Zio, E. (2014). Foundational issues in risk assessment and risk management. *Risk Analysis*, 34(7), 1164–72.

Baldwin, R., Cave, M., & Lodge, M. (2010). *The Oxford Handbook of Regulation*. Oxford: Oxford University Press.

Baldwin, R., Cave, M., & Lodge, M. (2012). *Understanding Regulation: Theory, Strategy, and Practice*. Oxford: Oxford University Press.

Batalden, B.-M., & Sydnes, A. K. (2014a). Maritime safety and the ISM code: A study of investigated casualties and incidents. *WMU Journal of Maritime Affairs*, 1(13), 3–25.

Batalden, B.-M., & Sydnes, A. K. (2014b). Risk assessments, key shipboard operations and soft barriers in offshore operations. *Proceedings of the European Safety and Reliability Conference, ESREL 2013, Safety, Reliability and Risk Analysis: Beyond the Horizon* (pp. 1611–18). Boca Raton, FL: CRC Press.

Batalden, B.-M., & Sydnes, A. K. (2015). Auditing in the maritime industry: A case study of the offshore support vessel segment. *Safety Science Monitor*, 19(1), Article 3.

Benn, J., Koutantji, M., Wallace, L., Spurgeon, P., Rejman, M., Healey, A., & Vincent, C. (2009). Feedback from incident reporting: Information and action to improve patient safety. *Quality and Safety in Health Care*, 18(1), 11–21.

Bhattacharya, S. (2012). The effectiveness of the ISM Code: A qualitative enquiry. *Marine Policy*, 36(2), 528–35.

Borch, O. J., & Batalden, B. (2014). Business process management in high-turbulence environments: The case of the offshore vessel industry. *Maritime Policy and Management*, 42(5), 481–98.

Bray, D. J. (2008). *DP Operator's Handbook: A Practical Guide*. London: Nautical Institute.

Castaneda, D., & Rios, M. F. (2007). From individual learning to organizational learning. Paper presented at the ECKM2007, Proceedings of the 8th European Conference on Knowledge Management, Barcelona, 6–7 September.

Crossan, M. M., Lane, H. W., & White, R. E. (1999). An organizational learning framework: From intuition to institution. *Academy of Management Review*, 24(3), 522–37.

Degani, A., & Wiener, E. (1990). *The Human Factors of Flight Deck Checklists: The Normal Checklist*. NASA Contractor Report, 177549. California, CA: Ames Research Center, Moffett Field.

Department of Transport (1987). *MV Herald of Free Enterprise: Report of Court No. 8074: Formal Investigation*. London: HMSO.

Granerud, R. L., & Rocha, R. S. (2011). Organisational learning and continuous improvement of health and safety in certified manufacturers. *Safety Science*, 49(7), 1030–9.

Grote, G. (2015). Promoting safety by increasing uncertainty: Implications for risk management. *Safety Science*, 71, 71–9.

Hadjimanolis, A., & Boustras, G. (2013). Health and safety policies and work attitudes in Cypriot companies. *Safety Science,* 52, 50–6.

Hale, A., & Borys, D. (2013). Working to rule, or working safely? Part 1: A state of the art review. *Safety Science,* 55, 207–21.

Heijari, J., & Tapanainen, U. (2010). *Efficiency of the ISM Code in Finnish Shipping Companies.* Turku, Finland: Centre for Maritime Studies.

Hollnagel, E. (2009). *The ETTO Principle: Efficiency-Thoroughness Trade-Off: Why Things that Go Right Sometimes Go Wrong.* Farnham: Ashgate Publishing.

Hollnagel, E. (2012). *FRAM, the Functional Resonance Analysis Method: Modelling Complex Socio-Technical Systems.* Farnham: Ashgate Publishing.

Hollnagel, E. (2014). *Safety-I and Safety-II: The Past and Future of Safety Management.* Farnham: Ashgate Publishing.

Huber, G. P. (1991). Organizational learning: The contributing processes and the literatures. *Organization Science,* 2(1), 88–115.

International Maritime Organization (1982). *Good Management Practice in Safe Ship Operation.* London: International Maritime Organization.

International Maritime Organization (1993). *Resolution A.741(18) International Management Code for the Safe Operation of Ships and for Pollution Prevention (International Safety Management (ISM) Code).* London: International Maritime Organization.

International Maritime Organization (2005). *Role of the Human Element: Assessment of the Impact and Effectiveness of the ISM Code.* London: International Maritime Organization.

International Maritime Organization (2010). *ISM Code, International Safety Management Code with Guidelines for its Implementation.* London: International Maritime Organization.

International Maritime Organization (2011). *STCW including 2010 Manila Amendments: STCW Convention and STCW Code: International Convention on Standards of Training, Certification and Watchkeeping for Seafarers.* London: International Maritime Organization.

International Maritime Organization (2014). *ISM Code, International Safety Management Code with Guidelines for its Implementation.* London: International Maritime Organization.

Jooma, Z., Hutchings, J., Hoagland, E., & Jandrell, I. R. (2016). The analysis of an incident investigation system. *IEEE Transactions on Industry Applications,* 52(6), 5235–40.

Kjellén, U. (2000). *Prevention of Accidents through Experience Feedback.* Boca Raton, FL: CRC Press.

Kongsvik, T. Ø., Størkersen, K., & Antonsen, S. (2014). The relationship between regulation, safety management systems and safety culture in the maritime industry. *Proceedings of the European Safety and Reliability Conference, ESREL 2013, Safety, Reliability and Risk Analysis: Beyond the Horizon* (pp. 467–73). Boca Raton, FL: CRC Press.

Lu, C.-S., & Yang, C.-S. (2010). Safety leadership and safety behavior in container terminal operations. *Safety Science,* 48(2), 123–34.

Marine Accident Investigation Branch (2016). *Report on the Investigation into the Listing, Flooding and Grounding of Hoegh Osaka Bramble Bank, The Solent, UK on 3 January 2015.* Southampton: Marine Accident Investigation Branch.

McDonald, N., Corrigan, S., Daly, C., & Cromie, S. (2000). Safety management systems and safety culture in aircraft maintenance organisations. *Safety Science,* 34(1), 151–76.

Mejia, M. (2001). Performance criteria for the International Safety Management (ISM) Code. Paper presented at the 2nd General Assembly of IAMU International Association of Maritime Universities, Kobe, 2–5 October.

Nonaka, I., Toyama, R., & Konno, N. (2000). SECI, Ba and leadership: A unified model of dynamic knowledge creation. *Long Range Planning,* 33(1), 5–34.

Oltedal, H. A. (2010). The use of safety management systems within the Norwegian tanker industry: Do they really improve safety? *Reliability, Risk and Safety: Theory and Applications,* 1–3, 2355–62.

Oltedal, H. A. (2011). Safety culture and safety management within the Norwegian-controlled shipping industry: State of art, interrelationships, and influencing factors. PhD thesis, University of Stavanger.

Ripamonti, S. C., & Scaratti, G. (2015). Safety learning, organizational contradictions and the dynamics of safety practice. *Journal of Workplace Learning,* 27(7), 530–60.

Stolzer, A. J., Halford, C. D., & Goglia, J. J. (2011). *Implementing Safety Management Systems in Aviation.* Farnham: Ashgate Publishing.

Tharaldsen, J. E., Wiig, S., Oestnes, H.-K., Ersdal, G., Hinderaker, R. H., Knudsen, S. et al. (2013). A system perspective on organisational learning. Paper presented at the European HSE Conference and Exhibition, London, 16–18 April.

Von Krogh, G. (1998). Care in knowledge creation. *California Management Review,* 40(3), 133–53.

Yang, C.-W., Fang, S.-C., & Lin, J. L. (2010). Organisational knowledge creation strategies: A conceptual framework. *International Journal of Information Management,* 30(3), 231–8.

4 Culture and maritime safety

Jon Ivar Håvold and Helle A. Oltedal

Introduction

Inquiries into accidents such as *Prestige*, *Herald of Free Enterprise*, *Sleipner*, *Scandinavian Star Estonia*, *Bow Mariner* and *Hoegh Osaka* ask questions about management systems and organizational culture, highlighting the importance of organizational and human factors as antecedents to accidents. The main purpose of this chapter is to provide students, managers and human factors practitioners with background on what culture can be and what safety culture research within shipping can teach us. This chapter reviews and discusses the culture construct and how cultural aspects can influence maritime safety.

The focus on safety culture took off when the term Safety Culture was used in the International Atomic Energy Agency's (IAEA) initial report (1986) following the Chernobyl disaster, resulting in the IAEA publishing a guide to safety culture in 1991. Many industries such as nuclear power production, aviation, hospitals, construction and chemicals started to focus on safety and culture during the 1990s (Håvold, 2005a). Traditionally, safety research within the maritime industry and shipping had its roots in natural science, addressing mainly technical issues. Research on culture and safety culture are relatively new in this setting, and started to catch the interest of researchers at the beginning of the third millennium (Berg, 2013; Hetherington, Flin, & Mearns, 2006).

As an indicator of research on safety culture in the maritime industry compared with other industries, a search on Google Scholar and ORIA was performed on 20 May 2017. The Google Scholar search listed 1,094 papers on nuclear safety culture, 598 on hospital safety culture, 9,180 on patient safety culture, 412 on construction safety culture, 281 on aviation safety culture, and only 154 on maritime safety culture. A search on ORIA using the terms 'safety culture', retrieving only peer-reviewed papers, listed a total of 334 on nuclear safety culture, 344 on hospitals, 1,424 on health, 193 on construction, 82 on aviation and 48 on maritime safety culture. A search on the database Science Direct using the terms 'safety culture' (title, abstract and key words) and 'maritime' (all fields), returned 42 results from journals and books, distributed annually as shown in Figure 4.1. The first peer-reviewed journal publication retrieved was from 2000.

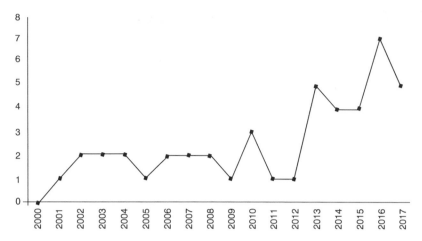

Figure 4.1 Number of safety culture publications in maritime domain (retrieved from Science Direct)

Although only three databases were searched using these broad terms, the results give a clear indication that there is room for more research on maritime safety culture; however, it is also clear that research interest in maritime safety culture is on the rise.

Some of the first studies that described social life and culture on board a modern vessel are from the late 1960s to the early 1970s; however, the cultural angle in these is more implicit, and safety issues are not captured (Eriksen, 1998; Roggema & Hammarstrøm, 1975; Roggema & Thorsrud, 1974). Unfortunately, many of these publications are in Norwegian only, although those without mastery of a Scandinavian language can find a summary of the research in the book: *Working on the Quality of Working Life: Developments in Europe*, chapter 11 (Johansen, 1979). Another interesting paper in this area is Lezaun, 2011, which describes and analyses the work of Johansen, Thorsrud and Emery to create a more democratic form of work organization aboard the Norwegian merchant ship *Balao* in the 1970s.

We can conclude from the statistics above that less safety culture research is carried out in the maritime domain compared with other industries, but that many similarities exist between all industries in terms of which antecedents/factors influence a 'good' or a 'bad' safety culture (Håvold, 2007a).

From a review of papers on safety culture in the maritime domain, the most important factor for safety culture seems to be positive attitudes to safety held by leaders both aboard and ashore (Ek, Runefors, & Borell, 2014; Håvold, 2005b, 2010b; Håvold & Nesset, 2009; Lappalainen & Salmi, 2009). Other important factors for safety culture appear to be satisfaction with safety activities/rules (Håvold & Nesset, 2009) and low work pressure (Ek et al., 2014; Håvold & Nesset, 2009). Antecedents to 'bad' safety culture seem to be fatalism, conflict between work and safety and obscure safety instructions (Håvold, 2005a; Håvold & Nesset, 2009).

It appears that the same safety culture factors/scales are important across national cultures. However, significant differences between nations and between regional clusters of nations exist (Håvold, 2005b, 2007a, 2007b; Håvold, Nesset, & Strand, 2011). Findings indicate that neighbouring nations might form regional cultures, such as a Northern Europe cluster or a South East Asia cluster (Hanges, Lyon, & Dorfman, 2005). The South East Asian culture is much more fatalistic than the North European culture, but research shows that Filipino officers are significantly less fatalistic than Filipino ratings, which could be attributed to the greater influence of organizational or professional culture on officers compared to ratings (Håvold et al., 2011).

Culture, culture and culture

The term culture has its roots in agriculture, i.e. when something is cultivated, rather than being due to nature. Anthropologist Edward Tyler is widely credited with the first 'modern' definition of culture in his book *Primitive Culture* (Tylor, 1920 (1871)): 'that complex whole which includes knowledge, belief, arts, morals, law, custom, and any other capabilities and habits acquired by man as a member of society'. Anthropology comes from the Greek meaning the study of humankind, and for a very long time cultural research by anthropologists focused on studying primitive groups of people in less developed countries. Pettigrew was the first scholar to use the term organizational culture (Pettigrew, 1979). In the 1980s, use of the concept organizational culture became very popular among progressive managers and consultants in business and was spread through books such as *Corporate Cultures* (Deal & Kennedy, 1982) and *In Search of Excellence* (Peters & Waterman, 1982). During the 1980s and 1990s the label 'culture' was applied to everything not fully understood in corporate life.

Culture is not an easy concept to define and when reviewing 'the culture literature' hundreds of definitions appear. Already in 1952, a critical review carried out by the American anthropologists, Kroeber and Kluckhohn, on the concept and definition of culture compiled a list of 164 different definitions (Spencer-Oatey, 2012). Anthropologists, psychologists, political scientists and sociologists all differ in their definitions of culture, along with the distinction between culture and climate, and the concepts of both culture and safety culture have, over time, been a theme of heated discussion, with little theoretical consensus emerging on the ontological, epistemological and methodological questions relating to the subject. The main differences in these questions seem to be: (1) What is the scope of safety culture and the relationship between culture and climate? (2) How does the concept relate to other organizational aspects and outcomes? (3) Which methods are most suitable for measurement? (Oltedal, 2011). Many researchers have already elaborated upon these fundamental questions and although we are well aware of the nature of the debate, this chapter will not be a new contribution to this; it rather discusses how the concept has been used within the maritime industry (Guldenmund, 2000; Havold, 2000). A more thorough discussion on safety climate is found in Chapter 6.

Another discussion, ongoing since the 1990s, is how safety culture is linked to organizational culture, professional culture and national culture. This debate has been carried out in both papers and books (e.g. Helmreich and Merritt (1998); Reason (1998)) and is continued under the heading 'Culture and safety in the maritime domain' within this chapter. Many scholars, among them Reason (1998), Cooper (2000) and Guldenmund (2000), view safety culture as a subculture in organizational culture, while Antonsen (2009) views safety culture as the way organizational culture is materialized when it comes to safety. Helmreich and Merritt (1998) proposed that culture fashions a complex framework of national, organizational and professional attitudes and values within which groups and individuals function. Their research shows a link between national cultures, organizational culture, stress and safety. In situations where national culture and organizational cultures are in harmony, fewer stress factors will influence safe behaviour; but in situations where national and organizational cultural values are in conflict, stress might influence safe behaviour and safe decision-making.

Within the myriad of definitions of culture that exist, one must make a choice. Thus, in this chapter we discuss further the most common cultural model that breaks the concept down to its central building blocks: basic assumptions, values, norms and artefacts as depicted in Figure 4.2. As an example in the upcoming discussion we use a well-known accident, the *Bow Mariner*, a chemical tanker that caught fire and exploded in February 2004, killing 21 of its 27 crew members. In Figure 4.3, additional relationships with safety are added and further discussed.

The *Bow Mariner* incident

On February 2004, the chemical tanker the *Bow Mariner* caught fire, exploded and sank east of Virginia, in the United States. Of the 27 crew members aboard, only six survived. The master, chief officer and chief engineer were Greek and the remaining officers and crew Filipino. This incident demonstrates many cultural perspectives that illustrate how national, ship and organizational culture might be among the underlying factors causing an incident, and culture is explicitly addressed in the report. The vessel was operated by the Greek operator Ceres Hellenic Shipping Enterprises Ltd. (hereinafter Ceres). Documentation showed that the *Bow Mariner* apparently fully complied with the ISM code; however, several indicators of non-compliance were found. For example, officers below the grade of chief officer and chief engineer, which in the case of the *Bow Mariner* comprised the entire Filipino crew, had not read applicable portions of the safety and quality manual. The main cause of the incident was incorrect inertion and cleaning of the vessel's cargo tanks. The Greek master ordered the crew to open all of the cargo tank hatches for the empty tanks once the vessel was at sea, but he failed to explain these instructions. The investigation report makes clear that the Filipino crew did not question the master's unsafe order due to fear of the Greek officers. The Ceres manual described the master's authority as follows:

The master has full authority over all persons (personnel and passengers) on-board his vessel. The master's authority is not questioned and must be supported and maintained by on-board personnel. Orders must be carried out and obeyed as said, in letter and in spirit. Refusal to do so is grounds for prompt disciplinary actions, including possible termination of employment.

Such absolute authority was not unusual aboard seagoing vessels at the time, and many would argue that such absolute authority is essential. However, within other high-risk industries it would seem awkward and a threat to safety (professional culture). The distinction between the Greek senior officers and Filipino crew was found to be remarkable, and the Filipino crew were given almost no responsibilities and were closely supervised in every task – a distinction we assume was based on nationality only (national culture).

This lack of trust was also in contrast to the content of the safety and quality manual. During the investigation, the US Coast Guard (USCG) found that this was normal practice on board all Ceres vessels and thus may be referred to as organizational culture (United States Coast Guard (USCG), 2005). As this incident touches upon professional, national and organizational culture, the case will be used to support our discussions and arguments throughout this chapter.

Elaboration of the cultural elements

Basic Assumptions are at the deepest level and are the core of organizational culture. These are assumptions and beliefs that will often be unconscious, but

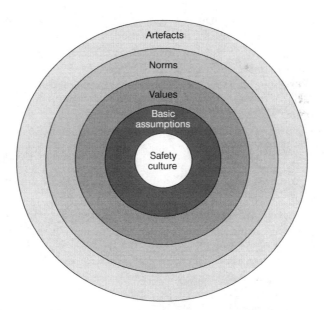

Figure 4.2 Generic model of safety culture

are taken for granted yet affect what is perceived and emphasized by people in the organization. Examples of basic assumptions are beliefs about human nature or how to define what is true and real. When one understands the underlying assumptions, it is easier to interpret the more observable and behavioural phenomena in which culture expresses itself (Schein, 1990).

These basic assumptions in a shipping company can be described as contained in a 'black box' which is locked and includes values that the group agree on, such as how the vessel and shipping company are governed, how problems are solved and situations treated. Basic assumptions are values used repetitively over a long period, eventually becoming truths for the organization or group. It seems that due to the way written rules, policies and procedures were practised, basic assumptions on the *Bow Mariner* were counter-productive to a 'good' safety culture and the officers (Greek) and the ratings (Filipino) had different and conflicting values. The *Bow Mariner* had an inert gas system so the empty tanks could be filled up with nitrogen to create a non-explosive atmosphere. In his procedures the Ceres' operator required tanks to be inert at all times except when being inspected or maintained. The requirement of inertion seems to have been ignored in the Ceres fleet, and only put in place when asked for by port state inspectors. It is unlikely that the chief officer performed any tests or issued any confined space entry permits even though entering a tank not tested for oxygen and that has not been gas freed is extremely dangerous.

Values (Schein, 1990) are important and lasting beliefs or ideals shared by the members of a culture about what is good or bad and desirable or undesirable. Values greatly influence a person's behaviour and attitude and serve as broad guidelines in all situations. All deeper levels of organizational culture begin as shared values and eventually become shared basic assumptions (over time after continuous use). If a master on a vessel believes that 'safety works' and convinces the crew to share their belief, and the crew through experience perceive that 'safety works', this will gradually be transformed into a shared value, and eventually into a shared assumption. Examples of other important values on a vessel are trust, honesty, integrity, cooperation, equality, respect, continuous improvement, etc. On the *Bow Mariner*, Ceres Hellenic Ship Enterprises Ltd. a Safety, Quality & Environmental Protection Management System (SQEMS) was established, documentation produced, audits completed and the documentation was in full compliance with the ISM code.

However, according to the investigation report after the accident, the Ceres SQEMS system was not functional aboard the *Bow Mariner*. This can be seen as an example of espoused values (Schein, 1990). Espoused values that are expressed on behalf of the organization or attributed to an organization by its senior managers in public statements may be different from practised values. It seems as though the practised values at *Bow Mariner* were far from the values expressed in the written documents. For example, the report stated that the three senior officers on the *Bow Mariner* had created a climate of fear and intimidation on the ship, violating important values for safety culture. Junior officers were prohibited from eating in the officers' mess, from reading the SQEMS documents, or from carrying out the jobs specified in them. Senior officers did not train their subordinates in the technical and administrative skills they needed to operate the vessel safely (International Maritime Organization (IMO), 2017).

Norms (Schein, 1990) are formal rules or standards laid down by legal, religious or social authorities against which the appropriateness (what is right or wrong) of an individual's behaviour is judged. The workgroup or team on a vessel defines what is socially appropriate and inappropriate based on values, beliefs, attitudes and behaviours and can judge and punish members who do not comply with these standards of behaviour. Norms are so important that failure to stick to the rules can result in severe punishment, the most feared of which is exclusion from the group. Other important norms aboard a vessel might be rules and regulations from the ship owner, rules and regulations from SOLAS, MARPOL, STCW, ISM code, Port State control, rules and regulations from the flag state, the classification authorities and from the insurance company, etc.

Artefacts (Schein, 1990) are physical, behavioural or verbal manifestations of an organization and its culture. Physical artefacts can be logos, buildings, dress codes or other material objects. Behavioural artefacts may include ceremonies, traditions or communication patterns. Verbal artefacts may include anecdotes, platitudes, stories or metaphors. Artefacts are the visible part of culture. A visible artefact in the merchant fleet is the hierarchical code for seafarers holding different ranks on ships. This ranking system claims to ensure smooth coordination of onboard operations and to promote proper management strategies. To employees in shipping company ceremonies, traditions, stories about the founder, the logo and colour of the vessel communicate belonging to the organization. Storytelling, dialogue and interaction are important in establishing and building organizational culture. Who are the heroes in the shipping company? Is the company hero a risk-taker who put the vessel and crew in grave danger, breaking the rules, but where, although close to a catastrophe, the situation ended well? Answers to questions like these say a lot about the safety culture in a company.

Further elaboration on safety culture

According to Bang (2013), organizations can have both one large overriding organizational culture as well as several subcultures simultaneously. In a shipping company, there seems to be a hierarchy of organizational cultures. First, there is an overriding organizational culture but each vessel will have its own subculture influenced by the overriding organizational culture; and on the vessel, there appear to be subcultures in the galley, engine room and on deck. A safety culture depends on recognition across the company, from management to workers, that the company has adopted safety and health as fundamental company values.

Although many definitions of safety culture have been used, many appear to have several themes in common:

- Safety culture is a concept defined at group level or higher, which refers to the shared values among all the group or members of the organization.

- Safety culture is concerned with informal safety issues in an organization, and is closely related to, but not restricted to, the management and supervisory systems.
- Safety culture emphasizes the contribution from everyone at every level of an organization.
- The safety culture of an organization has an impact on its members' behaviour at work.
- Safety culture is usually reflected in the contingency between reward system and safety performance.
- Safety culture is reflected in an organization's willingness to develop and learn from errors, incidents, and accidents.
- Safety culture is relatively enduring, stable and resistant to change.

A safety culture is related to and reflected in activities that have importance for operational safety such as communication, crew involvement, development of procedures and instructions, staffing and work pressure (see Figure 4.3).

The antecedents to safety culture, as illustrated in Figure 4.3, are discussed below using *Bow Mariner* as an example.

Procedures and safety rules, such as instructions for tank cleaning, were included in the cargo and ballast operations manual and other Ceres circulars. The SQEMS also cites Dr Verway's tank cleaning guide and tanker safety guide. The actual practice leading to the explosion was not in accordance with the operations manual and circulars. This is a clear example of espoused values where the values expressed in public statements differ from those shown in practice.

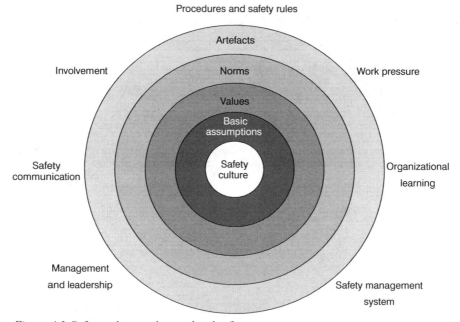

Figure 4.3 Safety culture and operational safety

Work pressure: 'None of the survivors went through indoctrination or familiarization as described in section 2.1.2 in the Fleet Operation Procedures Manual (FOPM) stating there was no time' (USCG, 2005, p. 28). In addition, lack of fire and boat drills (required by SOLAS) are an indication of work pressure and that safety had low priority. The fact that the *Bow Mariner* was two days behind schedule when arriving in New York also added to work pressure and stress.

Organizational learning is a process in which managers and employees within an organization learn to deal with new situations and problems so the organization becomes more skilled and 'experienced' (Argyris & Schön, 1996). The organization improves over time as it gains experience. The investigation report suggests that the organizational learning process did not function on the *Bow Mariner*.

A *Safety Management System (SMS)* is required by the ISM code. Every shipping company should develop, implement and maintain a SMS. Ceres established a SQEMS in accordance with the ISM code. As mentioned above there are numerous indicators showing that the SMS system was neither fully implemented nor functional aboard the *Bow Mariner*, leading to the conclusion that the SMS system was not functioning.

Training was scheduled and recorded in the Minutes of Safety Meetings but not conducted. Monthly fire and boat drills were not conducted as required. An overlap between crew as required in the FOPM was not met for new officers on their first voyage on the *Bow Mariner*. No familiarization training was conducted for new crew.

Management and leadership: It seems as though the leading officers both in the deck and engine departments had created a culture of lack of trust, based on fear. The subordinates were requested to do as they were ordered and not to ask questions. The management was abusive and threatened to send crew home if they did not work harder or questioned orders.

Safety communication was a problem on the *Bow Mariner*. In fact not only safety communication appeared to be a problem, but also all kinds of communication as the Greek officers communicated mainly in Greek, even though the official language was English. The crew were waiting for someone (the management) to tell them what to do. Just before the accident, the captain ignored questions from the Filipino third officer about the risks. Lack of technical knowledge, due to insufficient training and familiarization, and fear of the senior officers explain why the crew did not question the master's unsafe order to open all the empty tanks. The report says that the crew either did not know about the danger or were not inclined to question the master's order.

Involvement: On the *Bow Mariner*, the crew were not asked for their opinions (they were inclined not to question the master's orders) and management seemed to pressure the ratings to work faster even if conditions were unsafe. A crew that feels that the management is not serious about safety, violating safety rules on a grand scale and seeming only to care about productivity, will become disengaged (Maritime Accident Casebook, 2017).

Culture and safety in the maritime domain

As mentioned earlier in this chapter, while maritime safety culture is under-researched compared to safety culture research in other domains such as healthcare and the nuclear, construction and aviation industries, there have been some maritime safety culture research results reported. Much of the research on maritime safety culture has been carried out in the Nordic countries.

The maritime industry introduced the notion of safety culture quite early and in 1981, the International Chamber of Shipping (ICS) and the International Shipping Federation (ISF) published their Code of Good Management Practice (IMO, 2017), which advocated a 'culture of self-regulation of safety'. By the mid-1980s, the IMO drew attention to the role of human factors in maritime accidents and started to discuss what later became the ISM code. According to IMO, The ISM code is largely derived from the ICS/ISF Code of Practice, reflecting the development of the shipping industry's understanding of safety culture and safety management, with the acknowledgement that poor safety culture could be a reason for accidents and incidents. Until the adoption of the ISM Code, the IMO had attempted to improve safety in shipping largely by improving the hardware of shipping (e.g. the construction of ships and their equipment). By comparison, the ISM code focuses on the way shipping companies are managed, providing regulations and guidelines to promote the development of sound management and operating practices, and to foster the development of a sound safety culture (Oltedal, 2011).

Lappalainen and Salmi (2009) and Lappalainen (2008) came to the conclusion that the application of the ISM has improved maritime safety considerably. Kongsvik, Størkersen, and Antonsen (2014) interviewed maritime regulators, ship owners and seafarers on high-speed crafts to investigate the link between the ISM code and maritime safety culture. They found that the Safety Management System (SMS) appeared to contribute to a bureaucratic culture. The findings indicate that the structural measures resulting from the ISM code have led to a rule-oriented safety culture. However, the ISM code has instigated a more systematic approach towards safety, and a more holistic line of thinking about accidents (safety awareness) and their background causes.

Does organization matter?

Organizational culture matters. Cultures differs among ship owners and within the shipping company; each vessel has its own unique organizational culture (Håvold, 2007a). Within the vessel, different workgroups have different cultures; however, the relationship between the cultures at different organizational levels in the same organization is normally strong. An organizational culture influences almost everything and everyone. Leaders who develop a favourable organizational culture might find it easier to recruit the best people and keep them in the organization, while also having lower rates of absenteeism and better retention generally. The employees will be more motivated; more satisfied, more involved and be more responsive to organizational change. As mentioned earlier, many scholars such as Reason (1998) and Guldenmund (2000) view safety culture as a subculture of organizational culture where the key to safer organizations is trusting, open relationships. In a safe work culture, people speak up about unsafe situations, they do not stand silent when someone violates safe practices, they constantly look for ways to improve safety and they take personal responsibility for creating and maintaining a safe workplace. The values in organizations developing a favourable organizational culture are: safety for crew and public, security protection for ship and cargo, integrity for the reputation of the shipping company and continuous improvement where accidents and near accidents are viewed as learning opportunities. Håvold (2007a) benchmarked 16 ship owners and 142 vessels. He found significant differences among all the factors he used to measure safety culture both between ship owners and between vessels within the same organization. The factors found to most influence a positive safety culture in the maritime industry are the commitment of top management, the perception of officers' attitude to safety, satisfaction with safety activities/rules, a process of continuous improvement, motivated and empowered personnel on board and low work pressure. A literature review by Lappalainen (2008) showed that the role of top management is essential when establishing a safety culture and implementing a safety management system.

The factors found to influence negative safety behaviour most were fatalism, the perception of obscure safety instructions and a conflict between work and safety (Håvold, 2010b; Håvold & Nesset, 2009). Job satisfaction/dissatisfaction have the potential to both positively and negatively affect safety behaviour (Håvold & Nesset, 2009). The seafarer's occupation also had an impact on safety-related attitudes. Research indicates two clusters. Masters, deck and engine officers reported more positive safety attitudes/culture than galley officers, engine, galley and deck crew (Håvold, 2005b, 2010b). Age also seems to have an influence on safety-related attitudes. Younger seafarers reported significantly less favourable attitudes towards safety, less knowledge and higher work pressure than older seafarers (Håvold, 2010b). Research on Norwegian fishermen found that the older fishermen reported a higher level of stress but less conflict between work and safety than younger fishermen (Håvold, 2010a).

Does nationality matter?

National culture differentiates the national characteristics and value systems of particular nations. National culture influences organizational culture and professional culture as well as safety culture both directly and indirectly through training, organizational and professional culture (Håvold, 2007b; Helmreich & Merritt, 1998). In today's global economy, shipping employs crew from a wide variety of national cultures. Two-thirds of the world's merchant fleet have crews that are multinational and multilingual (Berg, Storgård, & Lappalainen, 2013; Horck, 2005; Magramo & Gellada, 2009).

According to Ajzen (1991), crew members with different backgrounds are likely to have different basic assumptions, values and norms, which in turn might influence their attitudes and behaviour (Gibson, 2004; Seo, 2005). Values and beliefs are also very much influenced by religion. People of different nationalities differ, for example, in their response to authority, how they deal with uncertainty and ambiguity, how they express their individuality, how they view time and space, and how they use gestures and eye contact. However, what goes for nationalities might also be the case among different subgroups within a country; it must be kept in mind that the term 'national culture' can be misleading. It may only refer to some of the people in a given country.

Hall (1976) introduced the concept of high context and low context to explain how people communicate in different cultures. High context implies that a lot of unspoken information is implicitly transferred during communication. People in a high context culture such as many countries in South East Asia and Latin America such as the Philippines, China, India, Indonesia, Brazil and Venezuela tend to place greater importance on long-term relationships and loyalty, and implement fewer rules and less structure. Low context implies that much information is exchanged explicitly through the message itself and rarely is anything implicit or hidden. People in low context cultures such as those of Northern Europe and North America, for example, Norway, Germany, the UK and USA, tend to accept short-term relationships, follow rules and standards closely and are generally very task-oriented. In a maritime context, this means that sailors must be aware that misunderstandings can occur during exchanges of information between sailors from low context and high context cultures.

Hofstede (1980) conducted the most referred-to study of how values in the workplace are influenced by national culture while also giving the shortest definitions of culture in general: 'Culture is software of the mind'. In 1980, he identified four dimensions of culture based on the results of a large-scale survey conducted within IBM; later two more dimensions were added:

Power Distance: The extent to which the less powerful members of a society accept that power is distributed unequally.

Individualism versus Collectivism: People look after themselves and their immediate family only versus people who belong to an in-group who look after them in exchange for loyalty.

Masculinity versus Femininity: Achievement and success versus caring for others and quality of life.

Uncertainty Avoidance: The extent to which people feel threatened by uncertainty and ambiguity and try to avoid such situations.

Long-term Orientation: The extent to which a society exhibits a pragmatic future-oriented perspective rather than a conventional historical or short-term point of view (added by Bond and Pang (1991)).

Indulgence versus Restraint: The extent to which a society allows free gratification of basic and natural human drives related to enjoying life and having fun, versus a society that suppresses the gratification of basic needs and regulates these by means of strict social norms (added by Minkov and Hofstede (2012)).

Power distance and uncertainty avoidance were considered by Merritt (2000) to be the two most important safety factors, something also supported by Lu, Hsu, and Lee (2016). In high power distance cultures, it is uncommon for ratings to question officers. Ratings expect detailed instructions (rule orientation) from superiors and generally accept them without reservations. They might follow what the officer says even if they know it is wrong. (This might partly be the reason for what happened on the *Bow Mariner* even if much more seemed to be wrong on that vessel.) In low power distance cultures individuals like to know what is expected from them, but want to decide how to do the work themselves (high degree of autonomy). In a maritime setting this means that officers managing crew from high power distance countries should focus on both what they should do and how to do it, but in low power distance countries managers should focus only on what to do if they trust the person.

Low uncertainty avoidance cultures are comfortable with uncertainty and relaxed in unplanned situations, in contrast to high uncertainty avoidance cultures who rely on and are comfortable with rule-oriented mechanisms for uncertainty mitigation. In a masculine culture, employees expect rapid promotion, non-routine work, high earnings and recognition from superiors. In a feminine culture, employees expect a pleasant workplace atmosphere, good relationships with colleagues and superiors as well as stable employment. In collectivistic cultures, employees normally do not look for challenges in their job and do not need the freedom to decide how to tackle them. Employees from individualistic cultures normally look for challenges at work and want to decide by themselves how to tackle them. This means that different types of management/leadership seem to function best in different cultures; however, one must be careful, as management is as much an art as a science. Individuals within cultures differ regarding religion, geography, sex, upbringing, education, etc. The above might also be very important for officers when it comes to how to motivate members of the crew. They will certainly not all be motivated by the same things (Deresky, 2014; Håvold, 2007b).

Hofstede's study has been criticized by several scholars, mainly because of the simplified and uncritical use of the factors but also because of the built-in western bias in the sample and questionnaire. Other criticisms are that the only unit of analysis was the nation state, and that the main research was performed in one company consisting of office workers only (Baskerville, 2003; McSweeney, 2002; Williamson, 2002).

Professional culture differentiates the characteristics and value systems of particular professional groups, i.e. the typical behaviour of deck officers or engine officers. Through personnel selection, education and training, on-the-job experience, peer pressure, etc., professionals tend to adopt the value system and develop behaviour patterns consistent with their peers; they learn to 'walk and talk' alike. This professional culture (and organizational culture) might be the reason for the large number of similar attitudes and values among Norwegian and Filipino officers sailing on Norwegian vessels (Håvold et al., 2011). They generally share a pride in their profession and are motivated to excel in it. On the other hand, they may adopt value systems that lead to the development of a sense of personal invulnerability, a feeling that performance is not affected by personal problems, or that errors will not be made in situations of high stress.

Regarding the *Bow Mariner*, it is natural to assume that the group of Greek officers and the group of Filipino crew differed with regard to what was shared, which also brings about the notion of subcultures. In general, it is assumed that subcultures are found within the shipboard departments – deck, engine and galley. Analyses of other levels in the organization may give different results, as the group may be defined as the whole fleet or as all shipping companies (Oltedal, 2011).

Where safety culture and organizational culture meet national culture

McGill Professor, Nancy Adler, discusses in her book *International Dimensions of Organizational Behavior* (2008) which culture is strongest – organizational culture or national culture. Her conclusion is that cultural differences were significantly greater among managers working within the same multinational corporation than they were among managers working for different companies in their own native country. When working for multinational companies, British seemingly became more British, Americans more American, Japanese more Japanese, and so on. According to Adler, the reasons are not well understood, but it appears that employees may be resisting a company's corporate culture if it is counter to the beliefs of their own national culture. Adler's observations support the conclusion that national culture outweighs organizational culture. The above research might indicate that influences of national cultures shape strong value systems among their members. The resulting shared values, preferences, and behaviours of population groups might differ widely between countries within the same company or between employees from different countries within the same work location. Håvold's (2007b) research indicates that vessels with crew from one country and from three or more countries reported a better 'safety culture' than vessels with crew from two countries. The findings seem to support Adler since an explanation might be that if crew from two nationalities work aboard a vessel, it is easier to polarize the value systems based on nationality, something that seems less probable if the crew hails from one culture, or from three or more cultures.

Håvold (2007a) found the main influential factors in safety culture were organizational culture and national culture. National cultures discriminate significantly on reported safety attitudes (Håvold, 2005b, 2007b). There seem to be national clusters, as illustrated in a survey on seafarers from six nationalities (Håvold, 2010b) where the seafarers appeared to form three groups: an Eastern European cluster with Poland and Latvia; a North Western European cluster with Norway and the Netherlands and a South East Asian cluster with Indonesia and the Philippines. The larger the cultural distance between two countries, the greater the uncertainty there will be at the cultural interface (Merritt & Maurino, 2004).

Håvold et al. (2011) also compared sailing officers from the Philippines and Norway, finding the two officer groups more similar than would be expected from previous research. The same factors influence positive safety attitudes for both groups. Fatalism is one of the factors that influence safety culture negatively (Håvold, 2007b) since people with a high degree of fatalism believe that what happens is an act of God, and therefore they cannot increase safety by training or doing something actively. This raises the question of whether Filipino officers are less fatalistic because of a strong professional culture, or a strong organizational culture. A person can be a member of many cultures and have many roles. An officer's loyalty can be to his country, to the shipping company, to his peers in the profession or to his colleagues. Soeters and Boer (2000) suggest that safety was enhanced when national culture was more individualistic in nature. They also noted that the greater the degree of collectivism in the culture of a country, the greater the predisposition towards regulation and the greater the degree of 'power distance' (Hofstede, 2001), which led to an increased likelihood of accidents.

In previous studies, there are divergent views as to whether multicultural crews pose a risk to maritime safety or actually improve safety. In the authors' opinion, it would be discriminatory to say more safety risks are posed if a ship has a multicultural crew or certain nationalities aboard. Rather the focus should be on how multiculturalism should be taken into consideration in ship operations, how good intercultural communication can be supported and how common safety culture can be implemented aboard ships with different national cultures.

Concluding remarks

Culture is a diverse and broad subject working simultaneously at the group, organizational, and national levels. Research on safety culture in general suggests that a higher level of safety culture measures was associated with more positive safety performance (Smith & Wadsworth, 2009). In this chapter, culture and culture in a maritime context have been discussed. Scant research on maritime safety culture is available before the year 2000 (Havold, 2000), and in the first scientific paper on maritime safety culture almost all reviewed research results come from the aviation, nuclear, chemical or construction industries. Later research on maritime safety culture (Håvold, 2005b; Håvold and Nesset, 2007) confirmed that safety measure scales seem to be relatively stable across risky industries including the maritime industry. Seafarers are members of many groups and many cultures. The same person can be a member of one or several work groups or teams, a department, a vessel, a shipping organization and several other organizations, a nation or a family. The many roles and cultures can clash, resulting in stress and fatigue, which in turn influences occupational health and safety.

Since maritime safety culture research is 'under researched', there are openings for many interesting projects that could increase our competence, enable us to take better decisions, and deliver education that is more focused and practical. Maritime safety culture should be investigated in more detail to see how health, safety and environmental issues can be prioritized in relation to each other and to other organizational goals such as efficiency and productivity. It might also be valuable to examine how 'good seamanship' is defined in different places around the globe and what the underlying values and norms are. Further research including the multilevel nature of shipping would also be rewarding. Is it the ship owner (organization), the vessel, the work group or the individual level that influences safety culture the most?

References

Adler, N. J., & Gundersen, A. (2008). *International Dimensions of Organizational Behavior* (5th edn). London: Thomson Higher Education.

Ajzen, I. (1991). The theory of planned behavior. *Organizational Behavior and Human Decision Processes,* 50(2), 179–211.

Antonsen, S. (2009). *Safety Culture: Theory, Method and Improvement.* Farnham: Ashgate.

Argyris, C., & Schön, D. (1996) *Organisational Learning II: Theory, Method and Practice.* Reading, MA: Addison Wesley.

Bang, H. (2013). Organisasjonskultur: En begrepsavklaring. *Tidsskrift for norsk psykologforening,* 50(4), 326–36.

Baskerville, R. F. (2003). Hofstede never studied culture. *Accounting, Organizations and Society,* 28(1), 1–14.

Berg, H. P. (2013). Human factors and safety culture in maritime safety (revised). *TransNav: International Journal on Marine Navigation and Safety of Sea Transportation,* 7(3), 343–52.

Berg, N., Storgård, J., & Lappalainen, J. (2013). *The Impact of Ship Crews on Maritime Safety.* Publications of the Centre for Maritime Studies A 64. Turku: University of Turku.

Bond, M. H., & Pang, M. K. (1991). Trusting to the Tao: Chinese values and the re-centering of psychology. *Bulletin of the Hong Kong Psychological Society*, 26/27, 5–27.

Cooper, M. D. (2000). Towards a model of safety culture. *Safety Science,* 36(2), 111–36.

Deal, T. E., & Kennedy, A. A. (1982). *Corporate Cultures.* Cambridge, MA: Addison-Wesley.

Deresky, H. (2014). *International Management: Managing across Borders and Cultures* (8th edn). Upper Saddle River, NJ: Prentice Hall.

Ek, Å., Runefors, M., & Borell, J. (2014). Relationships between safety culture aspects: A work process to enable interpretation. *Marine Policy*, 44, 179–86.

Eriksen, E. (1998). *Sjømannen ombord og i land: En studie av skipets sosiale organisasjon.* Bergen: Norse Publications.

Guldenmund, F. W. (2000). The nature of safety culture: A review of theory and research. *Safety Science,* 34(1), 215–57.

Hall, E. T. (1976). *Beyond Culture.* Garden City, NY: Anchor Press.

Hanges, P. J., Lyon, J. S., & Dorfman, P. W. (2005). Managing a multinational team: Lessons from project GLOBE. *Managing Multinational Teams: Global Perspectives* (pp. 337–60). Bingley: Emerald Group Publishing.

Havold, J. I. (2000). Culture in maritime safety. *Maritime Policy and Management,* 27(1), 79–88.

Håvold, J. I. (2005a). Measuring occupational safety: From safety culture to safety orientation? *Policy and Practice in Health and Safety,* 3(1), 85–105.

Håvold, J. I. (2005b). Safety culture in a Norwegian shipping company. *Journal of Safety Research*, 36(5), 441–58.

Håvold, J. I. (2007a). From safety culture to safety orientation: Developing a tool to measure safety in shipping. PhD thesis, Norweigan University of Science and Technology, Trondheim, Norway.

Håvold, J. I. (2007b). National cultures and safety orientation: A study of seafarers working for Norwegian shipping companies. *Work and Stress*, 21(2), 173–95.

Håvold, J. I. (2010a). Safety culture aboard fishing vessels. *Safety Science*, 48(8), 1054–61.

Håvold, J. I. (2010b). Safety culture and safety management aboard tankers. *Reliability Engineering and System Safety*, 95(5), 511–19.

Håvold, J. I., & Nesset, E. (2009). From safety culture to safety orientation: Validation and simplification of a safety orientation scale using a sample of seafarers working for Norwegian ship owners. *Safety Science*, 47(3), 305–26.

Håvold, J. I., Nesset, E., & Strand, Ø. (2011). Safety attitudes and safety ambivalence among officers from the Philippines and Norway. *Safety Science Monitor*, 15(1), 1–15.

Helmreich, R. L., & Merritt, A. C. (1998). *Culture at Work in Aviation and Medicine: National, Organizational, and Professional Influences*. Aldershot: Ashgate.

Hetherington, C., Flin, R., & Mearns, K. (2006). Safety in shipping: The human element. *Journal of Safety Research*, 37(4), 401–11.

Hofstede, G. (1980). *Culture's Consequences*. Beverly Hills, CA: SAGE Publications.

Hofstede, G. (2001). *Culture's Consequences: Comparing Values, Behaviors, Institutions and Organizations across Nations*. Thousand Oaks, CA: SAGE Publications.

Horck, J. (2005). *Getting the Best from Multi-Cultural Manning*. Copenhagen: BIMCO GA.

International Atomic Energy Agency (1986). *The Chernobyl Accident INSAG-1: A Report by the International Nuclear Safety Advisory Group*. Vienna: International Atomic Energy Agency.

International Maritime Organization (2017). Safety culture. http://www.imo.org/en/OurWork/HumanElement/VisionPrinciplesGoals/Pages/Safety-Culture.aspx

Johansen, R. (1979). Democratizing work and social life on ships: A report from the experiment on board M.S. *Balao*. In H. van Beinum (ed.), *Working on the Quality of Working Life: Developments in Europe* (pp. 117–127). London: Martinous Nijhoff.

Kongsvik, T. Ø., Størkersen, K., & Antonsen, S. (2014). The relationship between regulation, safety management systems and safety culture in the maritime industry. *Safety, Reliability and Risk Analysis: Beyond the Horizon* (pp. 467–73). Leiden: CRC Press.

Lappalainen, J. (2008). *Transforming Maritime Safety Culture, Evaluation of the Impacts of the ISM Code on Maritime Safety Culture in Finland*. Turku: University of Turku.

Lappalainen, J., & Salmi, K. (2009). *Safety Culture and Maritime Personnel's Safety Attitudes: Interview Report*. Turku: University of Turku.

Lezaun, J. (2011). Offshore democracy: Launch and landfall of a socio-technical experiment. *Economy and Society*, 40(4), 553–81.

Lu, C.-S., Hsu, C.-N., & Lee, C.-H. (2016). The impact of seafarers' perceptions of national culture and leadership on safety attitude and safety behavior in dry bulk shipping. *International Journal of e-Navigation and Maritime Economy*, 4, 75–87.

Magramo, M., & Gellada, L. (2009). A noble profession called seafaring: The making of an officer. *TransNav: International Journal on Marine Navigation and Safety of Sea Transportation*, 3(4), 475–80.

Maritime Accident Casebook (2017). Toxic leader – toxic culture: The death of the *Bow Mariner*. http://maritimeaccident.org/tags/bow-mariner

McSweeney, B. (2002). Hofstede's model of national cultural differences and their consequences: A triumph of faith – a failure of analysis. *Human Relations*, 55(1), 89–118.

Merritt, A. (2000). Culture in the cockpit: Do Hofstede's dimensions replicate? *Journal of Cross-Cultural Psychology*, 31(3), 283–301.

Merritt, A., & Maurino, D. (2004). Cross-cultural factors in aviation safety. *Cultural Ergonomics* (pp. 147–81). Bingley: Emerald Group Publishing.

Minkov, M., & Hofstede, G. (2012). Hofstede's fifth dimension: New evidence from the World Values Survey. *Journal of Cross-Cultural Psychology,* 43(1), 3–14.

Oltedal, H. A. (2011). Safety culture and safety management within the Norwegian-controlled shipping industry: State of art, interrelationships, and influencing factors. PhD thesis, University of Stavanger.

Peters, T. J., & Waterman, R. H., (1982). *In Search of Excellence.* New York: Harper & Row.

Pettigrew, A. (1979). On studying organizational culture. *Administrative Science Quarterly,* 24(4), 570–81.

Reason, J. (1998). Achieving a safe culture: Theory and practice. *Work and Stress,* 12(3), 293–306.

Roggema, J., & Hammarstrøm, N. K. (1975). *Nye organisasjonsformer til sjøs: Høegh Multina-forsøket.* Oslo: Tanum-Norli.

Roggema, J., & Thorsrud, E. (1974). *Et skip i utvikling: Høegh Mistral-prosjektet.* Oslo: Johan Grundt Tanum Forlag.

Schein, E. H. (1990). Organizational culture. *American Psychologist,* 45(2), 570–81.

Smith, A. P., & Wadsworth, E. J. K. (2009). *Safety Culture, Advice and Performance: The Associations between Safety Culture and Safety Performance, Health and Wellbeing at an Individual Level, and Safety Culture, Competent Occupational Safety and Health Advice, and Safety Performance at a Corporate Level.* Wigston: IOSH.

Soeters, J. L., & Boer, P. C. (2000). Culture and flight safety in military aviation. *International Journal of Aviation Psychology,* 10(2), 111–33.

Spencer-Oatey, H. (2012). *What is Culture? A Compilation of Quotations.* Warwick: Global PAD Core Concepts.

Tylor, E. (1920 [1871]). *Primitive Culture.* New York: J. P. Putnam's Sons.

United States Coast Guard (2005). *Investigation into the Explosion and Sinking of the Chemical Tanker Bow Mariner in the Atlantic Ocean on February 28, 2004, with Loss of Life and Pollution.* Washington, DC: United States Coast Guard.

Williamson, D. (2002). Forward from a critique of Hofstede's model of national culture. *Human Relations,* 55(11), 1373–95.

5 The human contribution

Helle A. Oltedal and Margareta Lützhöft

What's the fuss about the human element?

The worldwide population of seafarers serving in international trade is estimated to be approximately 1,545,000 people from virtually every nationality. Worldwide, there are about 90,900 vessels, registered in over 150 nations, which carry about 90 per cent of the world's trade; thus, more than one and a half million seafarers are transporting goods for the benefit of the world's population of about 7.49 billion (United Nations, 2016).

It is difficult to imagine a fully functional society without the maritime industry – shipping has been crucial for our wellbeing for centuries and is a driving force for innovation and change in society. Shipping has also resulted in several serious accidents, most of which are attributed to weather conditions, shortfalls in ship design, human error or a combination of all of these. Human error is said to be the principal cause in all – or at least most – incidents and accidents at sea (Anderson, 2003; Rothblum, 2000; Wagenaar & Groeneweg, 1987). In summing up a number of maritime causalities studies, Rothblum (2000) found that between 75 and 96 per cent of marine causalities are due – at least in part – to some form of human error. More recent research indicates that human error was a major contributing factor in about 60 per cent of all shipping accidents worldwide that occurred between 1997 and 2012 (Butt, Johnson, Pike, Pryce-Roberts, & Vigar, 2013), suggesting a decrease when compared to Rothblum's (2000) findings. Yet human error still dominates with a considerable share of contributory factors, hence creating the perception that the human element is the weak link in the maritime system.

One might say that the most efficient way to improve maritime safety is to get rid of the human element entirely. However, the human contribution to the maritime industrial system is essential for its existence, and, even in a possible future where unmanned and autonomous vessels might be a reality, there is no getting around human involvement. Even if they are unmanned, such 'drone ships' would be remotely piloted by on-shore operator. 'The autonomous ship does not mean removing human beings entirely from the picture, as it is sometimes stated. Unmanned ships need to be monitored and controlled and this will require entirely new kinds of work roles, tasks, tools and environments', said Eija Kaasinen, Principal Scientist at VTT Technical Research Centre of Finland (Rolls-Royce, 2016).

When managing safety and risk, contemporary theories, principles and practices usually cover the interaction between human, technology and organization.[1] Kaasinen's statement illustrates how change in one of the elements demands changes in others – new technology requires new forms of legislation, regulations, organization, education and even more technological innovation. However, the human has the main role in all this. After all, it is the people working in the systems – alone or in a group – that are making the decisions at all levels, be it at a policy level regarding international regulations through to IMO matters (see Chapter 2) or at an operational level, where those working at sea, or in an on-shore command centre, make everyday decisions. It is difficult to imagine the existence of the shipping industry without the involvement of people and as long as people are involved within the system, the human element will still need to be considered, now and in the future.

Human element, human factors and human error

The human element, human factors and human error are different concepts; however, there seems to be no clear consensus of the definitions and the nature of their interrelationship. Thus, the concepts are often misinterpreted and used interchangeably as if they were the same. Figure 5.1 depicts their interrelationship in accordance with our understanding.

In any maritime work operation, be it in offshore operations or transportation of goods or persons, humans work within a given system where they relate to a given technology and organization. The human, the technology and the organization will always be present, intertwined and influencing each other. For example, the future introduction of remote-controlled and autonomous vessels

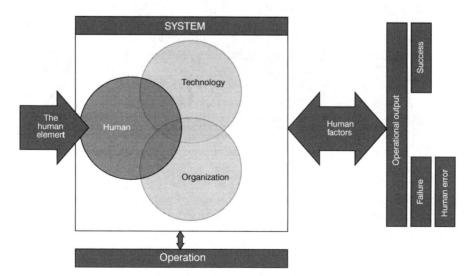

Figure 5.1 Human element, human factor and human error relationship

will demand other forms of technology (e.g. cyber security), organization (e.g. a shore-side command centre) and human competences required for navigation (e.g. from seafarer to 'drone-pilots'), which in turn will have consequences for safety, education, legal and ethical norms, among many others. The effects on safety management is discussed in Chapter 8.

In 1997, the IMO adopted the first resolution explicitly addressing the human element – Resolution A.850(20) (International Maritime Organization (IMO), 1997), which sets out the human element vision, principles and goals for the organization, and defines the concept that:

> the human element is a complex multi-dimensional issue that affects maritime safety and marine environmental protection. It involves the entire spectrum of human activities performed by ships' crews, shore based management, regulatory bodies, recognized organizations, shipyards, legislators, and other relevant parties, all of whom need to cooperate to address human element issues effectively …

The IMO definition is quite general and broad as it embraces every activity carried out by people. Another critique of this definition is that it narrows the scope of the output of human action to include only issues that affect safety, with the phrasing 'the human element is a complex multi-dimensional issue that affects maritime safety'. Following from this, we regard the definition to be reactive, as it only encompasses what is known to affect safety. In situations with a high degree of complexity and uncertainty, as we have in today's modern society, the output, whether the consequence or result of someone's actions, may be unknown beforehand, and thus needs to be experienced before safety measures are implemented.

The human element encompasses a wide range of different aspects, and there is no unified definition of the concept. Squire (2005) defines the concept as 'the term human element embraces anything that influences the interaction between a human and any other human or system or machine aboard ship'. The definition provided by Squire illustrates the conceptual confusion. The wording 'anything that influences the interaction' is normally used to describe human factors – not the human element.

The International Ergonomics Association defines human factors as ergonomics, which is the scientific discipline concerned with the understanding of interactions among humans and other elements of a system, and the profession that applies theory, principles, data and methods to design in order to optimize human well-being and overall system performance (International Ergonomics Association, 2012).

However, the term human factors is sometimes used synonymously with the human element. Although the wording is different, both definitions mention the *interaction* between the elements, and what *influences* that interaction – which in our view is the core of the definition of human factors.

The IMO definition does explicitly point out that the responsibility rests on all stakeholders in the maritime industry, and not only on the shipboard personnel.

These stakeholders may be ship owners, agents, classification societies, governmental bodies, the IMO, all of whom have different responsibilities in working towards safer shipping, e.g. development of regulations, implementation, execution and control. Given this, we find it even more intriguing that accident investigations still focus on errors made by sharp-end personnel, as with the *Costa Concordia* investigation, which will be further addressed later in this chapter. Although the human element relates to many groups of people, both inside and outside an organization, we believe that the seafarers themselves have the greatest interest in safety. After all, if something should go wrong, whatever the reason, the seafarers are the ones who risk losing their own lives or who bear the immediate responsibility for passenger safety. Therefore, we can assume that most errors made by crew members are unintentional and do not explicitly aim to cause an accident, but are induced by other factors.

In any given situation, the required operational output is efficient and safe operations, where the conditions for human error are minimized. Thus, human factors is concerned with creating acceptable working conditions for crew, a discipline concerned with designing machines, operations and work environments so that they match human capabilities, limitations, and needs, which is reflective of the relationship between people and the systems with which they interact. The focus should be to understand the factors that influence human performance – and how – with the aim of achieving a specific output, such as minimizing human errors.

About the human role in accidents

The indicated decrease of the human role in accidents might be a result of improved safety and better regulations; however, it could also reflect the common view of how the role of the human in an unwanted event is perceived. There have been several shifts in how investigators regard the role of the human in relation to other causal factors, from technological linear models to complex non-linear models, as reviewed in Chapter 1. The theoretical perspective on risk and safety has no doubt changed over the course of time, a change that is reflected in accident investigations and research, and there are also differences between countries and organizations. In an informal discussion with accident investigators from the Marine Accident Investigation Board Norway (AIBN) and the Swedish Accident Investigation Authority (SHK), it was evident that the two neighbouring countries had, at least at the time, two different approaches. In Sweden, it was customary that a human factors expert formed part of the investigation team from day one, whereas in Norway, it was customary that a human factors expert was brought in to the team *if* the appointed team during their investigation discovered that human factors were an issue. With the former approach, it is reasonable to assume that a team of investigators would discover and interpret human factors issues more correctly, whereas in the latter approach there is the possibility that human factors issues are not perceived or discovered at all.

However, while it is still common to refer to the numbers of human error, some analyses are up to three decades old and do not necessarily represent the cause(s) per se, but the causal opinion of the investigator and the accepted way of investigating unwanted events at the time of the investigation and the accident itself. An example illustrating this point is presented below: the foundering of MV *Lady Ann*.

On 16 September 1982, outside Western Australia, the offshore supply vessel *Lady Ann* assisted in the recovery of the anchors of the drilling ship *Regional Endeavour*. During the operation, the *Lady Ann* struck the *Regional Endeavour* on its fender, her bow then caught the handrails at the base of a crane and a life raft stand. These contacts damaged the *Lady Ann* enough for her to sink. The incident was investigated by the acting Senior Marine Surveyor and the six-page-long incident report is based on interviews with the crew (Australian Transport Safety Bureau (ATSB), 1982).

The incident report is typical for its time, identifying human error as the main cause, with no mention of issues such as organizational, underlying or latent factors. According to the sole investigator, the accident was caused by an error of judgement on the part of the master of the *Lady Ann* with the following explanation:

> He [the captain] had never attempted such a manoeuvre to a rig making way through the water before and seems to have misjudged his approach, possibly failing to appreciate the leeway the '*Regional Endeavour*' was making and probably going too fast. He did not appreciate and allow early enough for the interaction between the '*Regional Endeavour*' and his vessel. His action in stopping the port engine and using the bow thruster and starboard engine astern to attempt to straighten up was obviously instinctive and would probably have been correct had the '*Regional Endeavour*' not been making way. As the '*Regional Endeavour*' was moving it seems inappropriate ...

The final conclusion is 'the initial error is attributable to inexperience in the type of operation being attempted and perhaps overconfidence' (ATSB, 1982).

If this accident had been investigated today, organizational and latent factors would most likely have been analysed and the lack of experience in the type of operation that was attempted would today raise questions of the company's overall safety management system, which includes competence, skills and training, and why the company placed the captain in such an operation and situation. A company is responsible for keeping track of their employees' competence and for ensuring that their employees have proper training for the job to which they are assigned.

As mentioned, when addressing safety and risk within the maritime industry, it is common to refer to the considerably high number of cases of human error as a causal factor, without reflecting upon the weaknesses that are found behind how these numbers emerged. Since Wagenaar and Groeneweg's publication (1987), much has changed within the industry as well as in the theoretical frameworks; thus, hopefully, future research and publications will refer to human error as a historical occurrence, if referred to at all.

When considering the human element in relation to risk and safety, the focus has traditionally been on how humans have or could have erred, which is illustrated in the investigation of the foundering of *Lady Ann*. This approach towards safety – also called the person approach – focuses on the error of individuals such as forgetfulness, inattention, error of judgement or moral weakness (Reason, 2000). Reason distinguishes between two ways of viewing safety: the person approach and the system approach. It is pointed out that errors happen quite often and only a few, under certain conditions, lead to accidents. While the person approach focuses on all the possibilities for human fallibility, the system approach, on the other hand, concentrates more on the conditions under which people work, and focuses on building defences to prevent/manage errors or mitigate their effects. Under this view, human error is not seen as a direct cause, but as a consequence of weaknesses found elsewhere in the system, which are then considered root causes. The humans are still involved in the causal chain; however, the focus is shifted to other systemic factors that make up the workplace conditions.

Although contemporary safety management and accident investigation more frequently consider organizational factors, there is still a long way to go. National differences in theoretical perspectives and practices in investigations are illustrated with examples from the investigation of the *Costa Concordia* accident – an Italian cruise ship that capsized and sank in January 2012, resulting in 32 deaths, and the *Bourbon Dolphin* accident – a Norwegian anchor handling tug supply vessel that capsized and sank in April 2007, resulting in eight deaths.

The official *Costa Concordia* Investigation Report – published by the Italian Ministry of Infrastructures and Transports (2013) – states that:

> Therefore, distractions, errors and violations can be established as the elements which characterized the human factors as root causes in the *Costa Concordia* casualty.
>
> Those above mentioned recommendations have been made [report makes reference to stability and flooding, hull, vital equipment, emergency powering, redundancy of equipment, emergency management, minimum safe manning, muster list, and so on], despite the human element as the root cause in the *Costa Concordia* casualty.

Accident investigations are intended to identify the contributing latent and active factors, but too often result in a simple linear narrative that displaces more complex, and potentially fruitful, accounts of multiple and interacting contributions to how events really unfold (Peerally, Carr, Waring, & Dixon-Woods, 2017). We find it surprising that the human element (sometimes mistakenly referred to as human factors) is considered a root cause, since this view does not align with contemporary safety theories. There is no doubt that the crew on board made several mistakes, but we also argue that the vast majority of mistakes and errors are the result of systemic defects that need to be corrected. There are several challenges with root cause analysis, but when a root cause analysis is used correctly, it should detect systemic defects.

In the full *Costa Concordia* report, most errors, violations, slips and lapses are ascribed to the crew, whereas underlying organizational factors and the shipping company's responsibility are absent. As an example, we examine the communication problems due to language, identified in the investigation. The official working language on board was Italian, and all instructions and procedures were in Italian. The vessel was crewed with personnel of 38 different nationalities, and the investigation revealed that several of the crew members did not understand Italian orders properly or indeed, in some cases, not at all. In addition, procedures were not carried out in accordance with the Italian written instructions, which may be a result of crew not understanding the language. However, the company is not held responsible for staffing the vessel with crew that did not understand the official working language. In this case, the root cause should be considered to be the company's safety management system, not human error.

Others have also identified weaknesses and biases in the *Costa Concordia* investigation and note that, at the organisational level, latent conditions may be identified in various managerial processes, such as manning and human resources management, the acquisition of technology, the delivery of training, and – most critically – the engineering of a safety culture (Di Lieto, 2012). Di Lieto points to several underlying organizational factors that may have induced the accident. For instance, there is no doubt that the Bridge Resource Management (BRM) practices on the bridge were flawed, such as lack of team briefing, and lack of formal handover. However, it is unknown if these practices were regulated by company policies and procedures and/or supported by formal BRM training, which is the responsibility of the company. Di Lieto (2012) suggests that critical erroneous actions carried out by the crew, such as sail-pasts close to shore, were routine practice and thus should have been addressed in the company's formal safety management system.

In comparison, the factors leading up to the capsizing of the *Bourbon Dolphin* also involved many errors and procedural violations on behalf of the crew, including lack of communication and coordination. The incident was investigated by a Norwegian Commission appointed by Royal Decree (Norges offentlige utredninger, 2008), and leaves no doubt as to who is held responsible for the incident. In the key conclusions, mistakes made by the crew are toned down – focusing on the company's responsibility. This is illustrated here with two of the key conclusions:

Neither the company nor the operator ensured that sufficient time was made available for hand-over in the crew change.

The company did not make sufficient requirements for the crew's qualifications for demanding operations. The crew's lack of experience was not compensated for by the addition of experienced personnel.

It is also clear that the approach of the investigators is more in line with a contemporary theoretical perspective, as they state:

No chain is stronger than its weakest link. Where human beings are involved, experience shows that mistakes are made. That is why it is necessary to have safety systems that detect human error and make sure that it does not lead to accidents.

Human error can never be eliminated, and thus we need to understand how working conditions, workplace design, organizational systems and technology influence the seafarers during daily operations, how decisions are made and how they influence both risky and safe behaviour. In the following, we discuss how human element issues are perceived and how the industry addresses them in order to mitigate risk, using the perspective of human error as a direct (sharp-end) cause and organizational decisions as a latent causal (blunt end) factor.

From individual technical skills to team non-technical skills

The IMO International Convention on Standards of Training, Certification and Watchkeeping for Seafarers (STCW), 1978, was the first internationally agreed convention to address the issue of the human element through minimum standards of competence for seafarers. The STCW convention is the international instrument that sets forth the minimum knowledge requirements, seagoing experience and watchkeeping requirements for a ship's crew. Since then the seafarers' skills, training and competencies have been addressed by a complete revision of the STCW in 1995; further amendments in 2010, also referred to as the Manila Amendments, introduced BRM (Bridge Resource Management) training, shifting the focus from the individual to the bridge team. In addition, the development and adoption of the ISM code, addressed in more detail in Chapter 3, was introduced with the intention of providing a risk-based approach to managing shipboard safety and shifts responsibility away from the individual to the company (IMO, 1997).

After several serious marine casualities involving misjudgement and human error, the need for an increased focus on human-related activities in the safe operation of ships was acknowledged by the maritime community, and human element issues have been assigned high priority because of the prominent role of the human element in maritime casualties. While the STCW originally addressed the seafarers' technical skills, the Manila Amendments to the STCW Convention include a number of important changes, addressing non-technical skills such as situation awareness, decision-making, communication and coordination. After the STCW Manila Amendments, it became mandatory for ship officers at a management level to have competencies in leadership and managerial skills, e.g. workload management, resource management and decision-making (IMO, 2011). These skills are commonly taught in BRM training.

The BRM concept derives from the aviation industry, where it is called Crew Resource Management (CRM). In the aviation industry, the focus on human performance and its interaction with work systems started around the 1970s, triggered by serious aviation accidents and research on workload, errors, vigilance and decision-making (Smith, 1979). This led to a workshop,

'Resource Management on the Flightdeck', in the United States, sponsored by NASA and based on NASA research into air transport accidents. The focus of the workshop was the 'human error' in relation to interpersonal communications, decision-making and leadership (Helmreich, Merritt, & Wilhelm, 1999). An early definition of the concept was 'using all available resources – information, equipment and people – to achieve safe and efficient flight operations' (Lauber, 1986). The outcome of the workshop was the initiative for training of pilots to reduce unwanted outcomes by improved use of human resources. The evolution of CRM went through different stages, and the concept of CRM expanded continuously to also embrace 'the recognition and assessment of human factors issues' (Helmreich et al., 1999), organizational culture, automation, as well as the inclusion of other staff groups in aircraft operations.

In the maritime domain, the interest in human factors emerged in the period between WW2 and the end of the 1960s (Grech, Horberry, & Koester, 2008) with the start of 'Bridge Team Management' training on simulators which, although mainly technical, also encompassed some non-technical skills. However, unlike the aviation industry, the maritime industry did not recognize the need for a systematic CRM approach to training until the beginning of the 1990s, after several maritime accidents. BRM training courses were developed when seven major maritime industry bodies in collaboration with the Scandinavian Airlines System (SAS) Flight Academy decided to establish a global BRM training initiative (Hayward & Lowe, 2010). In particular, the grounding of the *QE2* at Martha's Vineyard, Massachusetts, in 1992 prompted the recommendation that pilots, masters and bridge personnel should have BRM training. At this time, training for maritime officers had traditionally focused on developing individual technical skills rather than addressing team management and non-technical skills such as decision-making, communication and coordination. The first BRM course was launched in 1993 (Kanki, Helmreich, & Ance, 2010). At that time BRM and human factors training was not mandatory, and until the Manila Amendments in 2010 it was only recommended that guidance be issued to masters and officers based on BRM principles. BRM was initially designed to improve the relationship between the master (ship's captain) and the pilot (a mariner who comes aboard to guide a ship through dangerous or congested waters), but soon transformed into the BRM course taught today, which addresses the broader implications of human factors for the safety and performance of mariners (O'Connor, 2011; Pekcan, Gatfield, & Barnett, 2005). Over time, team training and non-technical skills became regarded as important factors for safety.

In June 2006, the sub-committee on Flag State Implementation discussed the outcome of a report delivered by the Working Group on Casualty Analysis and Review of the STCW Code for the Investigation of Marine Casualties and Incidents, which identified that a high number of casualties was attributed to poor BRM practices, and stressed the importance of such training (IMO, 2007). The mentioned report (IMO, 2004) does not clearly convey what is meant by BRM practices, or to which principles the IMO is referring. After an extensive search, it was not possible to retrieve that information which appears to be well hidden within the IMO archives. According to Barnett, Gatfield, and Pekcan (2003), the concept of BRM is adapted directly from the aviation CRM model for training of non-technical skills, which raises the question of whether any sector-specific adjustments are carried out. Sector-specific characteristics may render an approach used in aviation less suitable for the maritime sector.

Aviation and maritime are both transportation systems, but that is the end of the similarity. For instance, within aviation, the crew working conditions are more fixed and it is possible to lead a normal life between shifts, pilots fly only the type of aircraft for which they are certified, the aircraft are standardized and all the technical equipment is placed within arm's reach. Furthermore, it is a highly proceduralized workplace (although shipping is certainly catching up) and pilots need to refresh their CRM training twice a year. In comparison, ship crew live in a more closed community; they both work and live within the same confined space for long periods. The deck officer certificate is neither ship nor ship type specific, although certain ships need special endorsements, such as tankers, and there is little standardization of equipment design and layout of the bridge. Furthermore, the BRM competence only needs to be demonstrated once in a lifetime.

The mandatory human element training requirements in the current STCW regulations are summarized in Table 5.1.

The required BRM training is mandatory for shipboard masters and officers, but not for all shipboard personnel nor shore-side staff. This may change when responsibilities for ship control move ashore (see discussion in Chapter 8). The optimal effects of BRM depend both on the operational and organizational levels, with the need for appraisal and commitment from the top management, as detailed in the ISM code, to support and ensure optimal working conditions. Most of the skills described in Table 5.1 rely on conditions provided by the shore side of the organization. For example, effective communication on board is challenging when crew members do not master the ship's official language, as in the *Costa Concordia* incident.

Table 5.1 Leadership and managerial skills in STCW

Leadership and managerial skills		
Task and workload management	*Resource management*	*Decision-making*
Planning and coordination	Allocation, assignment and prioritization of resources	Situation and risk assessment
Personnel assignment	Effective communication on board and ashore	Identify and generate options
Time and resource constraints	Decisions reflect consideration of team experiences	Selecting course of action
Prioritization	Assertiveness and leadership including motivation	Evaluation of outcome effectiveness
	Obtaining and maintaining situation awareness	

Team training – does it work at all?

BRM training is designed to mitigate human error. Research studies and maritime accident investigations list a multitude of concepts that supposedly can be addressed, such as lack of situation awareness, poor cooperation, poor planning, poor communication, poor teamwork, adverse workload, stress, fatigue, leadership and more (Akyuz, 2015; Baker & McCafferty, 2005; Chauvin, Lardjane, Morel, Clostermann, & Langard, 2013; Graziano, Teixeira, & Soares, 2016; Hetherington, Flin, & Mearns, 2006; Macrae, 2009; Mazaheri, Montewka, Nisula, & Kujala, 2015; Pekcan et al., 2005; Sandhåland, Oltedal, & Eid, 2015; Sandhåland, Oltedal, Hystad, & Eid, 2015; Uğurlu, Köse, Yıldırım, & Yüksekyıldız, 2015). Some of these studies explicitly mention deficient team training and BRM as causal factors, e.g. Chauvin et al. (2013) analysed 27 reports of collisions between vessels, 15 of which explicitly mention causal factors related to BRM, labelled as a precondition for human error. Macrae (2009) analysed 30 accident reports, which identified poor BRM as a significant causal factor in groundings. Uğurlu, Yıldırım, and Başar (2015) reviewed 131 maritime accident reports and attributed 101 causal factors to poor team management.

When BRM is identified as a causal factor, a common safety measure is to require ship officers to attend such training – this was the case with the *Costa Concordia* – although there is no research confirming whether BRM training within the maritime sector has had the intended effect such as within aviation. According to O'Connor (2011) the literature on the implementation of BRM training is far smaller than that on aviation CRM training. To our knowledge there are only a few research studies that have published results in peer-reviewed journals on the effect of BRM courses (O'Connor, 2011; Röttger, Vetter, & Kowalski, 2012; Röttger, Vetter, & Kowalski, 2015). There are reports on the effectiveness of BRM training which are not published in journals, and they will not be addressed in this chapter. The results of three of these unpublished studies are reported by O'Connor (2011).

O'Connor (2011) was the first to evaluate the effectiveness of the US Navy BRM program, measuring knowledge and attitude. Based on responses obtained from 142 navy surface war officers who had obtained BRM training, and 24 navy surface war officers who had not received BRM training, no significant difference between the two groups was found. Knowledge was measured by a ten-item multiple-choice test and attitude was measured by a 22-item questionnaire. O'Connor (2011) developed a version of the Flight Management Attitude Questionnaire (FMAQ), commonly used to measure pilot attitude and the skills taught in training and adapted for surface warfare officers, known as the Naval Aviator Human Factor (NAHF) questionnaire. The NAHF questionnaire consists of four factors:

1   My stress: emphasizes the consideration of, and possible compensation for stressors in oneself;
2   Stress of others: emphasizes the consideration of, and possible compensation for stressors in other team members;

3   Communication: encompasses communication of intent and plans, delegation of tasks and assignment of responsibilities, and the monitoring of team members; and

4   Command responsibility: appropriate leadership and its implications for the delegation of tasks and responsibilities.

No significant difference in the factor scores was found between the 142 Navy surface war officers who had obtained BRM training, and the 24 Navy surface war officers who had not received BRM training.

Röttger et al. (2012) studied the attitude and behaviour of two groups of participants from the German Navy: (1) 121 active seafarers who participated in ship bridge simulator training and (2) 101 junior officers who received a course in leadership studies. Both groups completed the Ship Management Attitudes Questionnaire–German Navy (SMAQ-GN) (also adapted from FMAQ) to measure attitude. The German version consisted of three scales: (1) Communication and Coordination (COCO), (2) Command Responsibility (COMMAND) reflecting a sense of shared responsibility, and (3) Recognition of Stressor Effects (RSE), regarding human performance in adverse conditions. Behaviour was assessed using the Non-Technical Skills behavioural observation system. The study confirms the existence of a relationship between attitude and performance, and that the relationship is non-linear where negative attitudes easily impede effective behaviour, but that behaviour is not as easily improved by positive attitudes. The study also suggests that SMAQ may be used as a benchmark in training needs assessment.

In a later study, Röttger et al. (2015) assessed the effectiveness of a classroom-based BRM training for junior naval officers. One hundred and seventeen junior naval officers participated in the study during their leadership training. Of these, 57 participants belonged to an experimental group and received BRM training. As in the 2012 study, attitude was measured with SMAQ-GN and knowledge by 13 open questions. Analyses of the questionnaire data did not reveal any differences between those who had received BRM training and those who had not. However, higher effectivity of BRM training as compared to the standard training was found on the level of reactions and learning criteria.

Although research has not established the effectiveness of BRM training, it does not necessarily mean that it is not effective, as all research has error margins. However, it would be plausible to believe these findings could be related to how the maritime community has approached BRM training, one example being the uncritical adoption of the aviation principles. Moreover, STCW does not define the skills needed, with the result that BRM training facilities have a large degree of freedom in how they want to design the syllabus and course content. From our experience of lecturing on safety management on BRM courses, we are left with the feeling that one size is assumed to fit all. Participants often come from various companies and vessels and one course class team can consist of captains of tugs, supply vessels and large construction vessels, so it is reasonable to question whether they receive equal benefit from the training, when the training situation is far from a real work setting.

We may also question the criteria that are provided in STCW for evaluation. For example, for knowledge and ability to apply decision-making techniques, the evaluation criteria are that decisions are the most effective for the situation. This vague phrasing leaves many open questions. Who decides what the most effective decision is, and in a real operation filled with uncertainty, is it even possible to reach the 'most effective decision'?

What is a good decision?

In the aftermath of an incident, poor decisions made by the crew are often highlighted, as was the case both in the *Costa Concordia* and the *Bourbon Dolphin*. Traditional views of decision-making focus on the normative, i.e. how humans should make decisions (if they were completely logical, in the way a computer is logical). However, research has diversified and we now use terms such as descriptive decision-making, where the focus is on how we actually make decisions. This type of research can be used to facilitate decision-making as opposed to telling people how they should do it. Normative views assume complete access to perfect data and enough time to evaluate and take appropriate action. This was critically examined by Dreyfus (1981), finding that people do not perform detailed comparisons of alternatives to make decisions but rather seem to use earlier experience to map against the situation at hand, and use pattern recognition to decide the course of action or response. The continued research into this is commonly referred to as recognition-primed decision-making, developed into naturalistic decision-making as described by Klein (2008).

The situations in which seafarers find themselves on-board are far from normative, perfect and predictable. We often hear that because ships move so slowly there is ample time to make decisions, as opposed to, for example, driving. However, it seems we use this time up in waiting for a situation to resolve itself, or be clarified. In the end, we have little time for action, and our safety margin has disappeared. What makes this even more difficult is that systems with feedback delays have a negative effect on human performance (Brehmer, 1992). Ships are examples of such systems, for example giving rudder orders and waiting for the effect. This is often misunderstood by those seeking to make decision support systems. There is research indicating that decision support might only be used when much time is available, i.e. routine situations and not when it would be most useful, in crises and emergencies (Grabowski & Sanborn, 2001).

In retrospective investigations, the investigators have all the time in the world to retrieve relevant information and analyse what would be the best decision in a given situation. However, when being *in* a situation, crew have neither all the relevant information, the time to analyse all the information nor the different decisional options in order to reach an optimal decision. They will make the best decision they can, given the circumstances and the conditions – using 'local rationality', which is described in more detail in Lützhöft and Dekker's discussion of the grounding of the *Royal Majesty* (2002).

In Korean waters in December 2007, the very large crude oil carrier (VLCC) *Hebei Spirit* collided with a crane barge. The weather conditions were poor. At the time of the incident, one of the towing wires of the crane barge being towed parted when the crane barge was in close vicinity of the *Hebei Spirit*. As a result, the crane barge made contact with *Hebei Spirit,* causing oil pollution to the sea area. Anti-pollution measures were immediately put into action on board the *Hebei Spirit* to reduce the spillage. The *Hebei Spirit* was laden with about 209,000 tonnes of crude oil, and due to the measures taken, an estimated amount of about 10,900 tonnes of cargo oil had spilled into the sea. The investigation conducted by the Marine Accident Investigation and Shipping Security Policy Branch revealed that the decision to commence the towing voyage when adverse weather had been forecast was the main contributory factor in this accident. Other causes of the accident were loss of control of the towing convoy in rough weather and delay in the notice to the vessel traffic information station and other vessels in the area by the tow Master (Maritime Accident Investigation and Shipping Security Policy Branch, 2007).

After the incident it was not only determined that the crew of *Hebei Spirit* were blameless, but also shown that the crew's actions demonstrated good seamanship and prevented a far larger oil spill into the sea (INTERTANKO, 2009). Thirteen days after the incident, the Korean Coast Guard completed their initial investigation concluding that Samsung, who owned and towed the barge, was only partially responsible for the collision, and only responsible for 10 per cent of the oil spilled. This was despite the fact that the cause of the accident was the breakage of the second-hand Samsung crane runner wire, which was never intended to be used for towing, between the lead tug and the barge. Both the captain and chief officer were detained in jail, and later the South Korean appeals court sentenced the *Hebei Spirit* captain to 18 months in jail and fined him 20 million won ($13,990). 'The captain could have averted a collision by pulling up the anchor or moving backward at full or half the usual speed', the appeals court said in its decision.

This case upset the maritime community worldwide, and several initiatives supported the captain and chief officer. The situation illustrates well how different groups have different opinions on what decision is the most effective for a situation. This takes us back to the criteria for evaluation of non-technical skills in STCW, not only 'decisions that are most effective for the situation', but also the example of 'Effective leadership behaviours are demonstrated'. Again, who decides what is the most effective leadership behaviour for a situation, and is it even possible to know?

Fatigue

Although the ISM code is a minimum requirement for shipping companies and masters managing shipboard safety, it should be noted that there are several safety factors that are not considered or pointed out clearly in the code. One of them is fatigue, which has attracted considerable attention from researchers and regulators in the maritime industry in recent years. There is, however, a subtle mention of it in the ISM code section 6, when talking about adequate resources. On an individual level, fatigue can greatly influence human health and wellbeing. It is now considered a shipboard hazard and therefore, from a workplace health and safety perspective, it is important for shipping companies in collaboration with on-board crew to focus on managing the risk of fatigue effectively.

As knowledge about fatigue and its effects on human performance has grown, there has been an increasing recognition of fatigue as a threat to maritime safety. A study conducted by the Marine Accident Investigation Branch (MAIB) showed that from 1995 to 2003, 'a third of all the groundings involved a fatigued officer alone on the bridge at night' (Marine Accident Investigation Branch, 2004). A later example is the *Shen Neng 1*, a bulk carrier that ran aground on the Great Barrier Reef in Australia on 4 April 2010. According to the investigation by the ATSB, a contributory factor in this accident was that the chief officer did not get sufficient sleep and was affected by fatigue. Another factor was that the shipping company did not implement effective fatigue management policies on board. This accident caused irreversible impacts to the maritime environment, which was regarded as the 'largest known direct impact on a coral reef by a ship grounding' (Christie, 2015).

In the maritime sector, the Maritime Safety Committee (MSC), at its 71st session (19 to 28 May 1999), considered the issue of fatigue and the direction in which the IMO efforts should be focused (IMO, 2001). Then the MSC approved the 'Guidelines on Fatigue' composed of several Modules at its 74th session, which were published on 12 June 2001. The Guidelines were aimed at assisting interested parties to better understand and manage fatigue (IMO, 2001). However, although deepening the understanding of fatigue, the Guidelines are now outdated and do not include the many lessons learned about fatigue management gained during recent years (Grech, 2016). Hence, Australia and other interested parties to the IMO initiated a process of reviewing and updating the Guidelines on Fatigue to incorporate a more holistic approach to managing the fatigue risks at sea (Grech, 2016). As a part of the review and update process, a maritime Fatigue Resource Management (FRM) framework with multiple defensive layers was proposed by Grech (2016, p. 6).:

As shown in Figure 5.2, the proposed maritime FRM framework is composed of two important processes, which contain five defensive layers to manage fatigue on board. The first process is FRM controls, dealing with layer 1 and 2 relating to control and management strategies for fatigue-related risks. The second process is FRM safety assurance, which includes layers 3, 4 and 5, providing data-driven feedback (assessment and evaluation) through monitoring to ensure the controls are working effectively. By combining those two processes, the maritime FRM will allow for continuous improvement and the identification of any areas in need of improvement. Furthermore, according to the proposed maritime FRM framework, the responsibility for fatigue management should be shared between shipping companies and seafarers. In order to support the inclusion of fatigue as a risk to be managed, we would also like to see fatigue and fatigue management

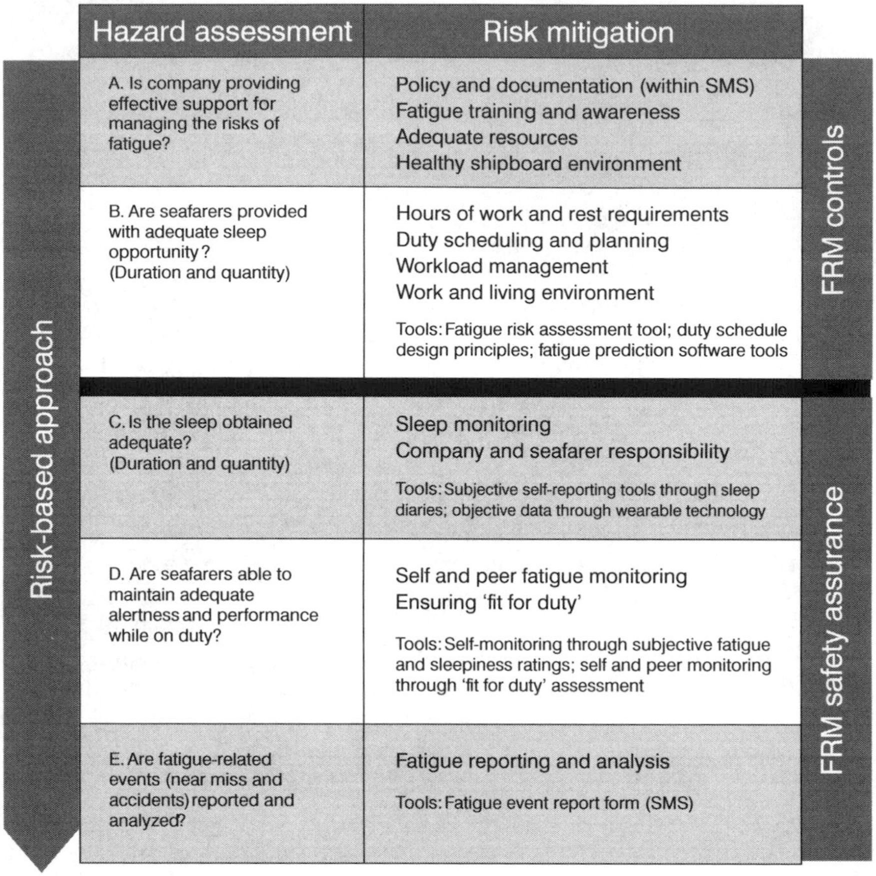

Hazard assessment	Risk mitigation
A. Is company providing effective support for managing the risks of fatigue?	Policy and documentation (within SMS) Fatigue training and awareness Adequate resources Healthy shipboard environment
B. Are seafarers provided with adequate sleep opportunity? (Duration and quantity)	Hours of work and rest requirements Duty scheduling and planning Workload management Work and living environment Tools: Fatigue risk assessment tool; duty schedule design principles; fatigue prediction software tools
C. Is the sleep obtained adequate? (Duration and quantity)	Sleep monitoring Company and seafarer responsibility Tools: Subjective self-reporting tools through sleep diaries; objective data through wearable technology
D. Are seafarers able to maintain adequate alertness and performance while on duty?	Self and peer fatigue monitoring Ensuring 'fit for duty' Tools: Self-monitoring through subjective fatigue and sleepiness ratings; self and peer monitoring through 'fit for duty' assessment
E. Are fatigue-related events (near miss and accidents) reported and analyzed?	Fatigue reporting and analysis Tools: Fatigue event report form (SMS)

(Left axis: Risk-based approach. Right axis: FRM controls; FRM safety assurance)

Figure 5.2 Proposed maritime FRM framework (Grech, 2016)

included more clearly than now in the STCW; it is mentioned in the Personal Safety and Social Responsibility skills but could be strengthened.

Another important aspect considered in the reviewed fatigue guidelines is the ship design aspect, ensuring that a human-centred approach is adopted in ensuring that ships are designed to optimize sleep quality (i.e. sleeping quarters location, orientation, comfort, etc.) when sleeping and alertness when working (ergonomic principles in the design of workspaces). Many fatigue-related issues such as undisturbed and comfortable sleep could be addressed by design. Design for safety is discussed in Chapter 7.

Looking into the future

There is no research that supports a claim that contemporary BRM team training has the desired effect, which might be due to lack of adaptation to the maritime industry. The sector-specific characteristics require, in our opinion, a different approach. We will draw attention to three areas for improvements. First, unlike aviation, vessels and maritime operations are less standardized, with a larger degree of variation and complexity. Despite this, most providers of BRM courses offer the same course to all participants, with the underlying assumption that one size fits all. An option would be to carry out team training on board the vessel itself, and let the instructor or assessor follow the bridge team and crew during real operations, and give feedback on their actual behaviour in real contexts. Second, while the STCW only requires ship officers and masters to undergo BRM training, safe operations and emergency preparedness is dependent upon good leadership of the entire crew, and communication and coordination also include those of lower ranks. Thus, we suggest that the current BRM training is expanded to include more of, or even the majority of, the crew, to become Crew Resource Management (CRM) training. In addition, shore-based key personnel should be required to undergo BRM training to ensure that shore-side decisions align with on-board needs and requirements. Third, there should be a requirement for retraining, as it is unrealistic to believe that attending a 20-hour course is sufficient to maintain the competences and skills needed for a lifelong occupation at sea, as suggested in the IMO model Course 1.39 Leadership and Teamwork.

Although 'human error' is still frequently mentioned in maritime incident reports, it is increasingly recognized in other sectors that such errors are the result of 'latent errors' performed long before, or far from, the sharp-end operations. In 2013, the International Labour Organization's (ILO) Maritime Labour Convention (MLC) entered into force. The MLC establishes minimum requirements for almost all aspects of working conditions for seafarers, and places a number of prescriptive requirements on ship and equipment design. However, the MLC also contains requirements relating to risk assessment and safe work and these would benefit from a human-centred approach. Human variability and adaptability pose challenges to both prescriptive and goal-based regulations.

Organizations and technology have changed a lot, and humans may have changed a little. Changes to date include:

- more complicated systems, with greater use of automation and remote technology;
- reduced manning scales, multi-cultural crews, evolving training standards;
- changing operating and commercial conditions; and
- a more demanding and complex regulatory environment.

In 2013, Rolls-Royce Holdings began developing unmanned ship technology, and in 2016 they completed a virtual reality command centre that serves as a prototype for the vision. This has spurred some groups to further develop autonomous technologies that in time may remove humans from the ship. Such 'drone ships' would be remotely monitored or controlled by on-shore captains, but all the on-board operations that crew members currently carry out, such as navigation and power management, would be handled by computer systems.

Advocates of autonomous technology argue that it will make shipping safer, less expensive and more environmentally friendly. The initiative and the claims have also been met with scepticism, however, and Chapter 8 discusses some related issues, challenges and changes.

Note

1 The terminology and abbreviations vary, most common is Human-Technology-Organization (HTO), Man-Technology-Organization (MTO), or Human-Organization-Technology (HOT)

References

Akyuz, E. (2015). A hybrid accident analysis method to assess potential navigational contingencies: The case of ship grounding. *Safety Science, 79*, 268–76.

Anderson, P. (2003). *Cracking the Code: The Relevance of the ISM Code and its Impact on Shipping Practices.* London: Nautical Institute.

Australian Transport Safety Bureau (1982). *Preliminary Investigation Carried out by W. B. Thomson into the Foundering of the M.V. 'Lady Ann' off Exmouth, Western Australia on 18th September 1982.* Australia: Australian Transport Safety Board.

Baker, C., & McCafferty, D. (2005). Accident database review of human element concerns: What do the results mean for classification? Paper presented at the Human Factors in Ship Design and Operation, London, 23–24 February.

Barnett, M., Gatfield, D., & Pekcan, C. (2003). A research agenda in maritime crew resource management. Paper presented at the Proceedings of the International Conference on Team Resource Management in the 21st Century.

Brehmer, B. (1992). Dynamic decision making: Human control of complex systems. *Acta Psychologica, 81*(3), 211–41.

Butt, N., Johnson, D., Pike, K., Pryce-Roberts, N., & Vigar, N. (2013). *15 Years of Shipping Accidents: A Review for WWF.* Southampton: Southampton Solent University.

Chauvin, C., Lardjane, S., Morel, G., Clostermann, J.-P., & Langard, B. (2013). Human and organisational factors in maritime accidents: Analysis of collisions at sea using the HFACS. *Accident Analysis and Prevention, 59*, 26–37.

Christie, T. (2015). Shipping hazards and protection of the marine environment: A second chance to properly value the world heritage listed Great Barrier Reef. https://www.environment-adr.com/uploads/WorldHeritage-GBR-Econ.Value-Shipping.8June15.pdf.

Di Lieto, A. (2012). *Bridge Resource Management: From the Costa Concordia to Navigation in the Digital Age.* Brisbane: Hydeas Pty Ltd.

Dreyfus, S. E. (1981). *Formal Models versus Human Situational Understanding: Inherent Limitations on the Modeling of Business Expertise.* Berkeley, CA: University of California.

Grabowski, M., & Sanborn, S. D. (2001). Evaluation of embedded intelligent real-time systems. *Decision Sciences, 32*(1), 95–124.

Graziano, A., Teixeira, A., & Soares, C. G. (2016). Classification of human errors in grounding and collision accidents using the TRACEr taxonomy. *Safety Science, 86*, 245–57.

Grech, M. (2016). Fatigue risk management: A maritime framework. *International Journal of Environmental Research and Public Health, 13*(2), 175.

Grech, M., Horberry, T., & Koester, T. (2008). *Human Factors in the Maritime Domain.* Boca Raton, FL: CRC Press.

Hayward, B. J., & Lowe, A. R. (2010). The migration of crew resource management training. In Barbara G. Kanki, Robert L. Helmreich and José Anca (eds), *Crew Resource Management* (pp. 317–44). San Diego, CA: Academic Press.

Helmreich, R. L., Merritt, A. C., & Wilhelm, J. A. (1999). The evolution of crew resource management training in commercial aviation. *International Journal of Aviation Psychology,* 9(1), 19–32.

Hetherington, C., Flin, R., & Mearns, K. (2006). Safety in shipping: The human element. *Journal of Safety Research,* 37(4), 401–11.

International Ergonomics Association (2012). What is ergonomics? http://www.iea.cc/ergonomics.

International Maritime Organization (1997). Resolution A.850(20) Human element: Vision, principles and goals for the organization. http://www.imo.org/en/OurWork/HumanElement/VisionPrinciplesGoals/Documents/850(20).pdf.

International Maritime Organization (2001). *MSC/Circ.1014 Guidance on Fatigue Mitigation and Management.* London: IMO.

International Maritime Organization (2004). *FSI 12/WP.2 Casualty Statistics and Investigation.* London: IMO.

International Maritime Organization (2007). *STW 39/7/7 Comprehensive Review of the STCW Convention and the STCW Code: Bridge Resource Management and Engine-Room Resource Management.* London: IMO.

International Maritime Organization (2011). *STCW Including 2010 Manila Amendments: STCW Convention and STCW Code: International Convention on Standards of Training, Certification and Watchkeeping for Seafarers.* London: IMO.

INTERTANKO (2009). Shipping world united behind Hebei Two. In *INTERTANKO Annual Review and Report 2008/2009* (pp. 32–34). Arlington, VA: INTERTANKO.

Kanki, B. G., Helmreich, R. L., & Ance, J. (2010). *Crew Resource Management.* San Diego, CA: Academic Press.

Klein, G. (2008). Naturalistic decision making. *Human Factors,* 50(3), 456–60.

Lauber, J. K. (1986). Cockpit resource management: Background and overview. In H. W. Orlady & H. C. Foushee (eds), *Cockpit Resource Management Training: Proceedings of NASA/MAC Workshop* (pp. 5–14). Moffett Field, CA: NASA Ames Research Center.

Lützhöft, M. H., & Dekker, S. W. (2002). On your watch: Automation on the bridge. *Journal of Navigation,* 55(1), 83–96.

Macrae, C. (2009). Human factors at sea: Common patterns of error in groundings and collisions. *Maritime Policy and Management,* 36(1), 21–38.

Marine Accident Investigation Branch (2004). *Bridge Watchkeeping Safety Study.* Southampton: Marine Accident Investigation Branch. https://www.gov.uk/government/uploads/system/uploads/attachment_data/file/377400/Bridge_watchkeeping_safety_study.pdf.

Maritime Accident Investigation and Shipping Security Policy Branch (2007). *Report of Investigation into the Collision between the Hong Kong Registered Ship 'Hebei Spirit' and Korean Crane Barge 'Samsung No. 1' on 7 December 2007.* Hong Kong: Marine Accident Investigation Section of the Hong Kong Special Administrative Region.

Mazaheri, A., Montewka, J., Nisula, J., & Kujala, P. (2015). Usability of accident and incident reports for evidence-based risk modeling: A case study on ship grounding reports. *Safety Science,* 76, 202–14.

Ministry of Infrastructures and Transports (2013). *Cruise Ship 'Costa Concordia' Marine Casualty on January 13, 2012.* http://www.cruisejunkie.com/Concordia%20report.pdf.

Norges offentlige utredninger (2008). *The Loss of the 'Bourbon Dolphin' on 12 April 2007.* Oslo: Norges offentlige utredninger.

O'Connor, P. (2011). Assessing the effectiveness of bridge resource management training. *International Journal of Aviation Psychology,* 21(4), 357–74.

Peerally, M. F., Carr, S., Waring, J., & Dixon-Woods, M. (2017). The problem with root cause analysis. *BMJ Quality and Safety* 26(5), 417–22.

Pekcan, C., Gatfield, D., & Barnett, M. (2005). Content and context: Understanding the complexities of human behaviour in ship operation. *Seaways: The Journal of the Nautical Institute,* p.14.

Reason, J. (2000). Human error: Models and management. *Western Journal of Medicine,* 172(6), 393.

Rolls-Royce (2016). Rolls-Royce reveals future shore control centre. https://www.rolls-royce.com/media/press-releases/yr-2016/pr-2016-03-22-rr-reveals-future-shore-control-centre.aspx.

Rothblum, A. (2000). Human error and marine safety. Paper presented at the National Safety Council Congress and Expo, Orlando, FL.

Röttger, S., Vetter, S., & Kowalski, J. T. (2012). Ship management attitudes and their relation to behavior and performance. *Human Factors,* 55(3), 659–71.

Röttger, S., Vetter, S., & Kowalski, J. T. (2015). Effects of a classroom-based bridge resource management training on knowledge, attitudes, behaviour and performance of junior naval officers. *WMU Journal of Maritime Affairs,* 15(1), 143–62.

Sandhåland, H., Oltedal, H., & Eid, J. (2015). Situation awareness in bridge operations: A study of collisions between attendant vessels and offshore facilities in the North Sea. *Safety Science,* 79, 277–85.

Sandhåland, H., Oltedal, H. A., Hystad, S. W., & Eid, J. (2015). Distributed situation awareness in complex collaborative systems: A field study of bridge operations on platform supply vessels. *Journal of Occupational and Organizational Psychology,* 88(2), 273–94.

Smith, H. R. (1979). *A Simulator Study of the Interaction of Pilot Workload with Errors, Vigilance, and Decisions.* NASA Technical Memorandum, 78482. California: Ames Research Center, Moffett Field.

Squire, D. (2003). Exploring the human element *The International Maritime Human Element Bulletin,* 1 (Oct 2003), 4–5.

Uğurlu, Ö., Köse, E., Yıldırım, U., & Yüksekyıldız, E. (2015). Marine accident analysis for collision and grounding in oil tanker using FTA method. *Maritime Policy and Management,* 42(2), 163–85.

Uğurlu, Ö., Yıldırım, U., & Başar, E. (2015). Analysis of grounding accidents caused by human error. *Journal of Marine Science and Technology,* 23(5), 748–60.

United Nations (2016). *Review of Maritime Transport 2016.* New York: United Nations.

Wagenaar, W. A., & Groeneweg, J. (1987). Accidents at sea: Multiple causes and impossible consequences. *International Journal of Man-Machine Studies,* 27(5–6), 587–98.

6 Risk perception

Michelle R. Grech

Introduction

Maritime accident investigations historically tended to focus on the behavioural factors or those aspects readily observable and directly attributable to the accidents under investigation. This meant that focus was limited to the sharp end – that is the seafarers – with lack of training, poor lookout, a disregard for safety rules, risk-taking behaviours, violations and complacency often cited as the causes. The 1989 capsize and sinking of the *Herald of Free Enterprise* is perhaps one of the first cases in which the investigation and subsequent inquiry went beyond the traditional 'individual' focus to uncover the deeper, organizational safety issues (Grech, Horberry, & Koester, 2008). This led to increased concern over the management of risk in shipping and henceforth the regulated implementation of safety management systems further discussed in Chapter 3.

The International Safety Management (ISM) Code was meant to address concerns over shipboard risks. However, the pervasive thinking that threats and hazards are contained within the ship and mainly emanate from the behaviours and actions of individual seafarers still exists. This means that in shipping the main focus of hazard identification and risk management and control is on actively identifying and preventing safety issues that reside only within the vessel's internal environment. This implies a large emphasis is on the 'sharp end' with a focus on the training of seafarers and their compliance with safety rules and procedures. This is evident even with the International Maritime Organization's response to major accidents, with the most recent example being the grounding and subsequent sinking of the *Costa Concordia* on 13 January 2012 (further discussed in Chapter 5). Apart from purely technical design issues, following this accident the IMO's main focus area was on updating passenger ship safety training, which is all well and good, if this were the only safety issue. This is endemic across the industry both on shore and at sea, and conflicts with current theoretical approaches being adopted which take a more systems approach to risk and safety as described in Chapter 5.

It is now well known that there is never only one thing wholly responsible for an accident. Accidents rarely happen from a single bad decision or action by the seafarer or pilot, but are often caused by multiple interacting factors. Often even normal, commonly accepted behaviours play a role in adverse events. This focus on the sharp end only, in effect, limits our understanding of the systematic issues that contribute to accidents (Reason, 2008), with such an approach leading to ineffective safety interventions being implemented. Issues such as poor design, shortfalls in procedures, working conditions, high work demands and job pressures are often overlooked, leaving the seafarer to develop informal 'high risk' practices and shortcuts over time to circumvent these deficiencies that are incompatible with the realities of daily operations. Some of these 'high risk' practices that seafarers adopt to 'carry on and do their job' can gradually become the norm and be perceived by the seafarer as normal and hence 'low risk'. As an example, on 20 December 2015, an able seaman (AB) on board a product tanker lost his life when the ladder he was climbing suddenly slipped and he fell on the main deck (Accident Investigation Board Norway (AIBN), 2017). The AB observed that the forward hook of the free-fall lifeboat was in need of lubrication. As the height from the deck to the forward hook was 4.8 metres above the main deck the AB asked the boatswain to help him keep a ladder in position below the lifeboat so he could reach the forward hook and complete the lubrication task. The ladder was not properly secured and was unstable as both feet did not have firm contact with the deck.

The AIBN (2017) found a number of factors that contributed to this accident; in particular, the work was carried out without a risk assessment or work permit, the procedure for work aloft was not complied with in this work situation, the AB was not wearing adequate personal protection equipment and no fall prevention measures were put in place. The AB and the boatswain who was holding the ladder in place for him at the time were both concerned with completing the preparation of the vessel before arrival in port, which on this particular voyage entailed a port state control inspection. They both wanted to finish the job, yet they chose to conduct a high-risk task without putting in place adequate control measures. So, why did a well-trained, highly experienced seafarer like the AB choose to conduct what we would normally view as a high-risk task without adequate control measures in place? The investigation report indicates that the AB would have been aware that lubrication of the forward hook for the free-fall lifeboat without a risk assessment and a work permit was a deviation from the procedures and the originally scheduled maintenance job. But is it possible that in this situation, 'informal practices' had already crept into the day-to-day work on this ship and for the AB and possibly other crew members this had become 'normal practice' and hence viewed as 'low risk'? In effect, this shows that this issue is more complex and requires a more systems approach to identify the less 'proximal' safety issues. This aspect was not discussed in the report but certainly warrants further study.

The research is now clear on the relationship between risk perception and behaviour (e.g. Rundmo, Nordfjærn, Hestad Iversen, Oltedal, & Jørgensen, 2011), indicating that subjective assessments regarding risk and safety in day-to-day operations may affect behaviour. Several studies on accidents and risk have demonstrated the value of taking into account perceptions of risk in the understanding of risk-taking in the explanation of accidents. Risk perception is often seen as the perceived likelihood that an individual will experience the effect of danger (Short, 1984). Indeed the way risk is viewed and perceived has been shown to influence safety behaviour (Cooper & Phillips, 2004). Basically, if perceptions of risk are faulty, risk management efforts are likely to be misdirected (Slovic, Fischhoff, & Lichtenstein, 1981). There has been increasing interest in the factors that shape perceptions of risk because of its influence on risk-taking behaviour (Rundmo et al., 2011), with some studies carried out within a maritime context (Bailey, Ellis, & Sampson, 2006, 2007, 2010). The need to understand risk perceptions in the maritime industry and their link to safety outcomes is important and will be discussed in the proceeding sections.

The concepts of 'safety' and 'risk'

Safety has recently been defined as 'the achievement and maintenance of the maximum intrinsic resistance to operational hazards' (Reason, 2008, p. 268). In other words, safety involves the ability to withstand hazards in day-to-day operations – usually achieved through the implementation of a combination of control measures. This is a shift from the traditional definition of 'freedom from accidents or losses' which points to an absolute absence from unwanted outcomes such as incidents and accidents. It is argued that this may not capture the reality of some industries such as maritime where the hazards – weather, crew errors, shipboard conditions, psychosocial factors and many more – are ever present, and the focus should be on what is meant to be 'safe'. To be 'safe' there must be control measures in place that enhance a system's resistance to operational hazards. Safety also needs to be maintained by making regular checks upon these control measures and safeguards. In effect, risk and safety are inextricably linked as in order to manage risks you need to implement safety measures through both reactive (e.g. analysis of incidents) and proactive processes (e.g. regular checks and evaluations on control measures and safeguards). Both risk and safety are assessed by taking into consideration the environment (i.e. the operational context).

In the field of engineering, risk is defined as 'the probability of an event occurring, that is viewed as undesirable, and an assessment of the expected harm from occurring' (Damodaran, 2007, p. 6). This definition aligns to most definitions of risk which according to most of the literature comprises two aspects – the first being the probability of an event and the second being the magnitude of the consequence. The International Organization for Standardization (2009) considers an event in the context of risk as a 'deviation from the expected' which can either be positive and/or negative. The concept of risk has matured to a level in which it is now measurable through an evaluation and assessment of both the likelihood of an event occurring and the consequence of that event. For instance, it is now possible to assess the risk of losing a container from a vessel, the risk of being electrocuted during maintenance work and the risk of falling from a height while working.

Contrary to quantitative and other formal measurements of risk (e.g. quantitative risk analysis, safety assessment, and successful operations), risk perception refers to each individual's subjective assessment, which may differ significantly from formal quantifiable measures. Risk perception refers to the subjective assessment of the probability of experiencing a negative event. As risk is often thought of as being 'objective' by legislators (and others alike), in reality however what is *acted upon* by the individual seafarer (or group of seafarers) is based on their perception of risk. The difficulty, which is often the case and will be discussed later in this chapter, is that the 'objective' (or calculated) risk does not match or correlate with the perceived risk. Hence, the understanding of risk perception in relation to maritime safety is of interest because it differs from 'objective' risks and has been shown to relate to safety and risk-taking behaviours.

Perceptions of risk

Risk perception is formally defined as the 'subjective assessment of the probability of a specified type of accident happening and how concerned we are with such an event' (Marek, Tangernes, & Hellesøy, 1985, p. 152). There is now a robust knowledge base on factors that influence the perception of risk. Traditionally, risk perception has been examined and treated as a cognitive process and was viewed as a form of deliberate reasoning through analytical systematic processing. However, over time it has come to be recognized that it is also highly dependent upon an individual person's emotions, which play a part in how risk is perceived. This perspective (Loewenstein, Weber, Hsee, & Welch, 2001) hypothesizes that responses to hazards and risky situations result in part from direct emotional influences which include both negative and positive emotions such as worry, fear, dread, anxiety and overconfidence. It has therefore been suggested that an individual's experience of risk can be separated into 'cognitive evaluation' and 'emotional' (or affective) components. For instance, the emotional judgement of severity of consequences has been found to be important in the perception of risk. As an example, the objective risk of being involved in a plane crash is statistically very low. Most people cognitively know that it is statistically very safe to fly; however, some may emphasize the potentially devastating consequences and perceive the risk to be high and this is due to their emotions regarding the possibility of an accident. Fear and dread of flying can also heighten risk perception. This supports the notion that emotional reactions to risky situations often diverge from cognitive assessments of those risks. When such divergence occurs, emotional reactions often drive behaviour. Rundmo (1999) argues that there are certain indicators in events that make people rate them as high or low risk. Fischhoff, Slovic, Lichteinstein, Read, and Combs (2000) suggested the following nine indicators as influencing the perception of risk:

- voluntariness of risk: whether people enter into risky situations voluntarily;
- immediacy of effect: to what extent is the risk of death immediate, or is death likely to occur at a later time;
- knowledge about the risk by the person who is exposed to the potentially hazardous risk source;

- knowledge about the risk in science;
- control over the risk;
- newness, i.e. are the risks new and novel or old and familiar;
- chronic/catastrophic, that is a risk that may kill people one at a time (chronic risk) or a risk that can kill a large number of people at once (catastrophic);
- common/dread, i.e. whether people have learned to live with and can think about the risk reasonably and calmly, or is it a risk of which people have great dread on the level of a gut reaction; and
- severity of consequences.

Several studies have found that 'control over risk' in particular plays a large role in the perception of risk with a negative association between perceived control and risk perception (e.g. Fischhoff et al., 2000). If an individual perceives themselves as being in control of a situation or an activity (e.g. manoeuvring a vessel), the perceived risk associated with that activity tends to be lower compared to an individual who feels less in control of an activity. For example, people experience more control while driving a car than being mere passengers on an air flight. This could explain some of the differences in risk perception related to these activities. A maritime example of this is the fatality of a seafarer who fell from a ladder during cleaning operations in the cargo hold of a bulk carrier. The seafarer fell from a height of about one metre, but because his helmet was not secured properly it came off during the fall, resulting in his head taking a direct impact which knocked him unconscious. The investigation report indicated that, prior to the fall, the seafarer had removed the safety harness in order to move it to the next cleaning area. The low height may have influenced the seafarer's false belief that it was safe to do so. The investigation report indicated that the seafarer's behaviour may have been influenced by his feelings of control over the situation thereby leading to a low perception of risk. He also wanted to get the job done and removing the safety harness provided him with unhindered access to descend the ladder (AIBN, 2014). This example matches social cognition models which subscribe to the notion that the probability of engaging in risk-taking behaviour increases if the risk is perceived to be low; similarly the opposite applies – where the risk is perceived as high, behaviours tend to be more risk averse.

Risk perception is now viewed as a multidimensional construct that involves a combination of an individual's cognitive evaluation of the likelihood of the unwanted event (e.g. injury, loss, accident, etc.) caused by exposure to a risk source, as well as emotions (e.g. fear, dread, worry, etc.) related to the event. An important aspect about risk perception is that it cannot be viewed in isolation and the social context in which the person resides must be considered when looking at their perception of risk and how it affects safety behaviour.

Perception of risk and safety in a maritime context

The literature points to wide variations in the perception of risk across individuals. Factors such as age, nationality, gender and experience have been shown to mediate risk perception, with some of these examined within a maritime context. The work by Bailey et al. (2006, 2007, 2010) focusing on the perception of risks with regards to maritime incidents and personal injuries revealed the identifications of subcultures based on nationality, group (shore-based or shipboard), rank, age, the type of ship on which the individual works, and sea-going experience. Data were collected from 2,372 seafarers (senior officers (n=709); junior officers (n=704); and ratings (n=763)) and 104 shore-based managers from 50 countries who were asked to consider their perceptions of risk relating to the likelihood of a maritime accident (e.g. fire, collision, explosion, sinking, grounding and contact) and personal injury occurring within their company. Subjective perceived risks of accidents were compared with objective risks obtained from maritime accidents statistical data (Bailey et al., 2010; Pomeroy & Earthy, 2017). As shown in the vertical axis in Figure 6.1 the most common type of accidents at sea in order of priority (based on the derived statistics) are: (1) collisions; followed by (2) groundings; (3) contact; (4) sinking; (5) fire; and (6) explosions. The perception of risk as determined by respondents for each of these events varied when compared with the objective risk through the use of statistical data. The diagonal solid line in Figure 6.1 represents the point at which the objective risk (from statistical data) matches the perceived risk. The stars in the figure represent the subjective perception of risk for each of the accident types. The black stars depict an overestimation of the subjective perception of risk when compared to the objective risk, and the other stars represent an underestimation of the subjective perception of risk when compared to the objective risk. Fire for example, was generally perceived to be a high-risk event, with participants tending to overestimate the subjective perception of the risk of fire when compared to the objective risk obtained from statistical data. Shipboard crew are continuously exposed through training, awareness, focusing on fire, fire-fighting and control and this could explain their heightened perception of the risk of fire. In addition, 'fear' as an emotional response to fire may play a role in how fire is perceived at sea. Fire creates a mental picture of burning, pain and death, producing this heightened perception of the risk of fire, more so than collisions and groundings which were perceived as lesser threats.

In general, most respondents underestimated the risk associated with a collision, grounding and sinking. For grounding and sinking, shipboard crew were more likely to underestimate the risk than shore-based staff. Interestingly, there were no differences in perception for risks associated with a collision and/ or grounding between navigational and engineering officers. Being in control of the ship possibly affected this low risk perception. Pomeroy and Earthy (2017) indicate that the mistaken belief of being in control is potentially a feature of new technology, with crew feeling 'more in control' and less likely to have an accident because of their level of confidence and trust in these systems, which they argue is higher than is justified. This aspect is further discussed in Chapter 7.

Differences across groupings were also identified when looking at the perception of the likelihood of injury. These were based on a list of 18 possible events related to personal injury which were provided to guide responses. The list included events such as *contact with moving machinery*; *exposure to fire; being hit by a moving object; slips, trips or falls* and others.[1]

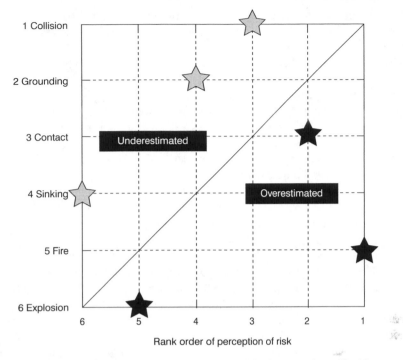

Figure 6.1 Perceived risks matched against objective risks for a number of incidents

When looking at the perception of risk of injury, integrated ratings identified 'working in a hot environment' as the highest risk leading to injury. Shore-based personnel and officers identified 'slips, trips and falls' and 'handling, lifting or carrying' as having the highest risk of injuries. According to the European Maritime Safety Agency's data, the latter estimation is closer to reality with 'slips, trips and falls' and 'handling, lifting and carrying' being the two most common accidents causing injuries on ships (European Maritime Safety Agency (EMSA), 2016). When risks are well known, the perceived risk to some extent will be a reflection of the 'real' risk. Potentially shore-based personnel and to some extent officers may have access to insurance claims related to injuries, providing them with the knowledge base to make a more accurate perception of these risks.

Bailey et al. (2010) also identified differences in perception of risk depending on the type of ship on which respondents worked. Different ships, and types of operation, have different characteristics which possibly influence perceptions of susceptibility to different types of events and consequential injuries occurring. For example, the risk of injury was perceived as lowest on tankers and perceived as highest on passenger vessels when compared to other types of vessels. Data derived from the EMSA (2016) show that a high number of occupational injuries occur on passenger ships and cargo ships and to a lesser extent on tankers. Although this may be a reflection of the data in terms of reported incidents and population by ship type, it nonetheless reflects seafarers' perception of injury by ship type.

Respondents perceived 'slips, trips and falls' and 'handling, lifting or carrying' as posing a greater risk of injury on passenger vessels when compared to all other vessel types. This could possibly be explained by the large number of work tasks associated with handling, lifting and carrying on these types of vessels. In effect, data show that the highest proportion of occupational injuries on passenger ships are associated with 'slips trip and falls', followed by 'handling equipment' (EMSA, 2016) showing that the perceptions conform to actual risks.

Those on tankers perceived the risk of injury as high in relation to the events 'exposure to harmful substances' and 'drowning/lack of oxygen/overcome by fumes'. Those working on bulk carriers perceived the risk of injuries resulting from 'being hit by a moving vehicle' and 'falls from height' as higher than other vessel types. In effect the EMSA (2016) data also indicate that 'slips, trips and falls', although not as high as on passenger ships are nonetheless the highest occurring occupational injuries on bulk carriers and tankers. Although priorities varied, most of the respondents perceived a medium to high risk of all these events occurring and leading to injury.

As indicated in Chapter 4, perception of risk is also known to be influenced by the social and cultural background that reflects the values and beliefs of people coming from different countries (Weinstein, 1980). To be more effective, intervention strategies for managing risk in the maritime sector need to also consider the fact that behaviours may vary between groups as they perceive risk differently. This is reflected in the studies by Bailey et al. (2007, 2010) in which nationality was identified as one of the most influencing factors in determining perception of risk of accidents and injuries. Bailey et al. (2007) included higher representative groupings from the Philippines, the United Kingdom, China, India and the Netherlands. The findings identified differences in perception of risk by nationality (see Figure 6.2), except for sinking and explosion which matched the overall population's perceptions for these events. Of interest is the difference in perception of the risk of fire between Chinese seafarers and all other nationalities. Most nationalities heavily overestimated the risk of fire, while Chinese seafarers more accurately reflected the data in their perception of the risk of fire.

The perception of the risk of injury type revealed a more clear separation in the perception of risk based on nationality (Bailey et al., 2010). The study findings revealed that respondents from the Philippines generally perceived the risks of injury as higher than any other national grouping – possibly indicating that seafarers from the Philippines in general perceive the maritime industry as a 'high' risk occupation, more than the other nationality groupings. Perception of risk may influence risk taking and approach towards safety concerns (Kouabenan, 1998) with findings from Bailey et al.'s (2010) study reflected in maritime injury data collected by nationality (Ádám, Rasmussen, Pedersen, & Jepsen, 2014; Hansen, Laursen, Frydberg, & Kristensen, 2008), suggesting that the risk of injury on board may vary by nationality. Hansen et al. (2008) found that seafarers from South East Asia, mainly the Philippines, may have a genuinely lower risk of occupational accidents compared with seafarers from Western and Eastern Europe. This is also reflected in more recent studies by Ádám et al. (2014) in which nationalities were

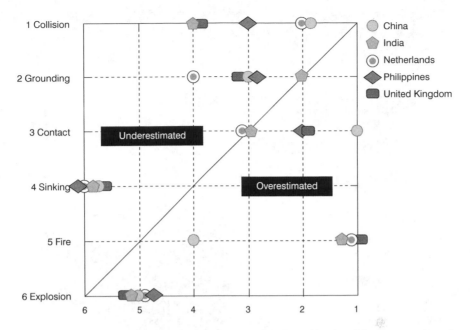

Figure 6.2 Perceived risks by nationality matched against objective risks for a number of incidents

divided into four groups comprising Western Europe (the majority from Denmark), Eastern Europe (the majority from Poland), South East Asia (the majority from the Philippines) and India which formed its own category. Ádám et al.'s (2014) findings showed low injury rates among South East Asian (Philippines) and Eastern European (Poland) seafarers when compared with Western European seafarers (Denmark). Bailey et al.'s (2010) findings showed that seafarers from the Philippines perceived the risk of injury as higher than other nationality groupings, possibly suggesting that they may be more risk averse, which is reflected in the lower injury rates amongst seafarers from the Philippines. Figure 6.3 provides a visual representation of how data on nationalities sit within the high–low risk perception and how this may influence injury rates. Of course, the reporting culture of various nationalities might skew this data. For example, the literature points to under-reporting of incidents from crew nationalities predominantly from the Philippines because of fear of being seen as underperforming possibly affecting career and job prospects. However, the data by Hansen et al. (2008) were adjusted for major confounders to provide a true comparative assessment, with the data still reflecting a significantly lower injury rate for seafarers from South East Asia when compared with seafarers from Western Europe.

Reasons for these differences could vary. For example, national working practices in which seafarers from Western European countries (i.e. Danish and Dutch) may be more compelled to work independently may mean they take more risks. High risk perceptions associated with seafarers from South East Asia may

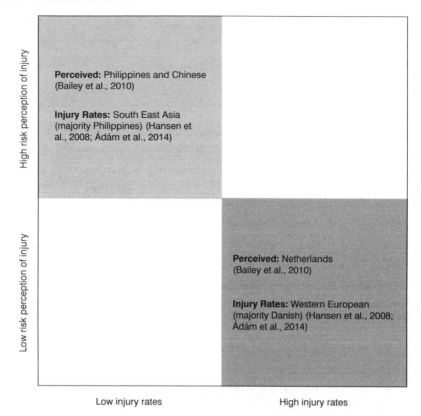

Figure 6.3 Risk perception and injury rates by nationality

be related to their national safety practices, with lack of adequate social security, employment (length of contract, job security, etc.) and labour market conditions compelling them to be more careful and afraid of being injured and subsequently losing their only family income. Higher estimates of risk are usually positively associated with precautionary behaviours. Although this is reflected in the data, it is problematic to conclude that this indicates more cautious behaviour among crew from the Philippines without further studies.

Safety climate and risk perception

Risk perceptions are also affected by the organizational safety climate, which is defined as 'employees' perceptions of their organization's policies, procedures and practices relating to safety' (Griffin & Neal, 2000). These perceptions provide a frame of reference, allowing, for example, seafarers to gauge the importance with which safety is regarded within their organization (Zohar, 2010). Safety climate differs somewhat from the concept of safety culture which is further discussed in Chapter 4. While safety culture is the value placed on safety and can be considered to be the 'personality' of the organization, safety climate is usually considered as the perceived value placed on safety at a particular point in time, more synonymous with the 'mood' of the organization (Yule, 2003). Safety climate can be considered as a snapshot of safety culture (Neal, Griffin, & Hart, 2000). The concept of safety climate is extensively covered in the literature and core to this is that it describes the worker's perception of their organization's commitment to safety (i.e. policies, procedures and practices relating to safety) which in turn influences workers' safety attitudes and behaviours (Neal et al., 2000; Zohar, 1980). Safety climate is closely associated to operations and is characterized by day-to-day perceptions towards the working environment, working practices, organizational policies and management. It is normally associated with the comprehension through which a group of people, such as seafarers, understand the hazards of the work environment around them. From a safety climate perspective, the environment, functioning, beliefs and safety values of the organization in which a seafarer works have been found to correspond to their perception of risk. The organizational safety measures and associated procedures in which these are implemented and executed determines how people working within that organization perceive the job resources available (such as shipboard climate and support from officers and other crew; role clarity; participation in decision-making; task significance, task identity; autonomy, etc.) as being effective enough to help them cope and deal with daily hazards (Bakker, Demerouti, & Verbeke, 2004). In effect, peoples' impression of job resources may constitute core aspects of what is considered to be the 'safety climate'. Thus, safety climate can, for those who share it, represent a natural and unquestionable way of behaving, and as such, it serves to elaborate a particular version of how risk is perceived.

An example that illustrates this point is that of a boatswain on board a bulk carrier who was crushed to death while working by one of the end stops as the jibs were being swung in (AIBN, 2009). The ship was making preparations to leave the port of Vancouver in Canada after taking on board a cargo of wood pulp destined for Japan. The boatswain and a deck cadet, together with an electrician who was operating the jibs at the time, were securing the ship's gantry cranes and holds when the accident happened. The company and crew confirmed that the boatswain who died in the accident was a highly experienced, responsible and conscientious seaman. He had been working for ten years with the same shipping company and had joined the vessel two months prior to the accident. Among other tasks, he was responsible for securing the cranes prior to sailing and making sure the deck crew used and operated deck equipment and machinery safely. The investigators identified that changes to the crane design over many years resulted in vulnerabilities within the system leading to what is typically referred to as a 'drift into failure'.

Reason (1990) refers to these as 'pathogens', which are pre-existing hazards that make the system vulnerable to failure. The newer updated designs of enclosing the top of the cranes led to these 'pathogens', which meant a continuous reduction in the visibility from the operator's cabin to the outside securing operation positions, thus eroding one of the safety control barriers of providing the crane operator with an overview of the outside environment. This meant that the electrician who was operating the jib at the time could not see the boatswain standing near the end stops. Risk and safety management practices were not revisited in line with the enclosed design changes and a consequent lack of adequate risk control measures in place resulted in the securing operations being carried out haphazardly. Despite the fact that the operation of securing the cranes was carried out by highly experienced personnel who had all undergone safety management system training, the system failed with consequential outcomes. It is possible that years of successful operations of securing the cranes may have provided the crew with the wrong perception that the risks were low. Hazardous activities for which personal risks are underestimated tend to be seen as under the individual's control and may be perceived as low risk (Fischhoff et al., 2000). The crew's perception that the requirement to reinforce the crane securing operation safety management system as low priority is reflected in their behaviour. It can be argued that long periods of success may breed complacency and indicates that wariness of success is important in high risk organizations (Hopkins, 2005). This reinforces the fact that knowledge about seafarers' risk perceptions and ratings of safety is necessary for the development of a safety climate in which each person accepts responsibility for working safely. It is also important to ensure that risk perceptions and 'real risk' are kept as aligned as possible through, for example, continuous safety training and ensuring procedures are kept up to date to reflect and reinforce safe practices.

Research has shown that the safety climate can influence risk perceptions and associated safety behaviours while performing work. If seafarers for example perceive their work environment as hazardous this affects their job satisfaction which negates specific job resources required to cope with the demands of their work (Bakker et al., 2004). Hence, consideration of the safety climate is important in a shipboard environment in which the work activities by their very nature pose a number of significant challenges to seafarers. Although it is evident that shipboard safety has improved substantially over the last decade, seafaring is still considered to be one of the most hazardous occupations, with fatal injuries significantly higher when compared to shore-based workplaces, and this situation is still evident in more recent studies (Roberts & Hansen, 2002; Roberts & Marlow, 2005; Roberts, Nielsen, Kotłowski, & Jaremin, 2014).

Safety climate has been shown to be a good indicator of safety performance in an organization (Griffin & Neal, 2000) and a useful tool in ascertaining workers' perceptions of the way that safety is being put in practice (Cooper & Phillips, 2004). The relationship between safety climate, safety behaviours and safety outcomes has also been identified as being reflected within the maritime domain (Lu & Tsai, 2008, 2010). In a study involving 291 seafarers from container vessels, safety climate was examined in relation to crew fatalities and vessel failures. Lu & Tsai (2008) found that the seafarers' risk perception was very much influenced by the number of fatalities and vessel failures. In particular, where the crew's views about aspects such as safety practices, safety attitudes and safety training were positive, this was related to fewer fatalities. Consistent with other studies in this area, Lu and Tsai (2010) also found a significant positive relationship between safety climate and self-reported safety behaviours in a study of 608 seafarers engaged on container vessels. Seafarers were more likely to report positive safety behaviours if they perceived safety rules and policies (safety management systems), management values and supervisor's (such as masters, chief mates and chief engineers) safety behaviours as effective. It has been argued that, if perceived risk affects behaviour, then influencing risk perception may result in behaviour change (Rundmo, 1999). In attempting to encourage positive behaviours, industries such as airlines and health care have in place a form of non-technical skills (NTS) training through what is commonly referred to as Crew Resource Management. CRM training focuses on the concept of 'error management' and has shown to positively affect safety behaviour. In the maritime sector, this is better known as Bridge Resource Management (BRM) and is discussed in more detail in Chapter 5.

Being 'afraid' is not so bad

It cannot be stated enough that placing a priority on safety, safety commitment and promotion within an organization is important for risk and accident protection behaviour (Rundmo, 1999; Rundmo et al., 2011). This is further supported by Griffin and Neal (2000) who found that perceptions of knowledge about safety and motivation to perform safely, influenced individual safety performance and also mediated the link between safety climate and safety performance. To influence perceptions of risk and safety, some highly reliable organizations reinforce what is referred to as 'chronic unease'. That is a continual belief that something will go wrong, creating continual risk aversion thinking. This equates to being 'on guard', 'alert' and possibly constantly 'afraid' of the possibility of imminent operational dangers. Due to the impossibility of devising a set of rules which adequately covers every situation, Hopkins (2005) identifies this as the most effective way to manage risk and safety. This still needs to be supported by a mature reporting system with the intention that data from near misses, failures and incidents are used to reinforce or slow down the inevitable process of 'forgetting to be afraid' (Reason, 2008).

Note

1   (1) Contact with moving machinery; (2) Being hit by moving (includes flying/falling) object; (3) Being hit by moving vehicle; (4) Being struck against something fixed or stationary; (5) Handling, lifting or carrying; (6) Slips, trips or falls on same level; (7) Falls from a height; (8) Trapped by something collapsing/overturning; (9) Drowning/lack of oxygen/overcome by fumes; (10) Exposure to, or contact with, a harmful substance; (11) Exposure to fire; (12) Exposure to an explosion; (13) Contact with hot surfaces; (14) Contact with cold surfaces; (15) Contact with electricity or electrical discharge; (16) Working in hot environment; (17) Working in cold environment; (18) Acts of violence

References

Accident Investigation Board Norway (2009). *Report on Investigation of Marine Accident MV Star Java: IMO no 9310513 Occupational Accident in Squamish 18 August 2008.* Lillestrøm: Accident Investigation Board Norway.

Accident Investigation Board Norway (2014). *Report on Marine Accident: MV Favorita LAGM6/9298519 Shipboard Occupational Accident in the East China Sea on 24 August 2013* (Marine 2014/10). Lillestrøm: Accident Investigation Board Norway.

Accident Investigation Board Norway (2017). *Report on Occupational Accident, Fall on Board Mariner (LAGQ7) on 21 December 2015 Off the Coast of California.* Lillestrøm: Accident Investigation Board Norway.

Ádám, B., Rasmussen, H. B., Pedersen, R. N. F., & Jepsen, J. R. (2014). Occupational accidents in the Danish merchant fleet and the nationality of seafarers. *Journal of Occupational Medicine and Toxicology,* 35(9), 1–8.

Bailey, N., Ellis, N., & Sampson, H. (2006). *Perceptions of Risk in the Maritime Industry: Ship Casualty.* Cardiff: SIRC.

Bailey, N., Ellis, N., & Sampson, H. (2007). *Perceptions of Risk in the Maritime Industry: Personal Injury.* Cardiff: SIRC.

Bailey, N., Ellis, N., & Sampson, H. (2010). *Safety and Perceptions of Risk: A Comparison between Respondent Perceptions and Recorded Accident Data.* Cardiff: SIRC.

Bakker, A. B., Demerouti, E., & Verbeke, W. (2004). Using the job demands-resources model to predict burnout and performance. *Human Resource Management,* 43(1), 83–104.

Cooper, M. D., & Phillips, R. A. (2004). Exploratory analysis of the safety climate and safety behavior relationship. *Journal of Safety Research,* 35(5), 497–512.

Damodaran, A. (2007). *Strategic Risk Taking: A Framework for Risk Management.* Englewood Cliffs, NJ: Prentice Hall.

European Maritime Safety Agency (2016). *Annual Overview of Marine Casualties and Incidents 2015.* Lisbon: European Maritime Safety Agency.

Fischhoff, B., Slovic, P., Lichteinstein, S., Read, S., & Combs, B. (2000). How safe is safe enough? A psychometric study of attitudes towards technological risks and benefits. In P. Slovic (ed.), *Perception of Risk* (pp. 80–104). New York: McGraw-Hill.

Grech, M. R., Horberry, T., & Koester, T. (2008). *Human Factors in the Maritime Domain.* Boca Raton, FL: CRC Press.

Griffin, M. A., & Neal, A. (2000). Perceptions of safety at work: A framework for linking safety climate to safety performance, knowledge, and motivation. *Journal of Occupational Health Psychology,* 5(3), 347–58.

Hansen, H. L., Laursen, L. H., Frydberg, M., & Kristensen, S. (2008). Major differences in rates of occupational accidents between different nationalities of seafarers. *International Maritime Health,* 59(1–4), 7–18.

Hopkins, A. (2005). *Safety, Culture and Risk: The Organisational Causes of Disasters.* Canberra: CCH.

International Organization for Standardization (2009). *ISO 31000:2009 Risk Management: Principles and Guidelines.* Geneva: International Organization for Standardization.

Kouabenan, R. D. (1998). Beliefs and the perception of risks and accidents. *Risk Analysis,* 18(3), 243–52.

Loewenstein, G. F., Weber, E. U., Hsee, C. K., & Welch, N. (2001). Risk as feelings. *Psychological Bulletin,* 127(2), 267–86.

Lu, C.-S., & Tsai, C.-L. (2008). The effects of safety climate on vessel accidents in the container shipping context. *Accident Analysis and Prevention,* 40, 594–601.

Lu, C.-S., & Tsai, C.-L. (2010). The effect of safety climate on seafarers' safety behaviors in container shipping. *Accident Analysis and Prevention,* 42(2010), 1999–2006.

Marek, J., Tangernes, B., & Hellesøy, O. H. (1985). Experience of risk and safety. In O. H. Hellesøy (ed.), *Work Environment Statfjord Field: Work Environment, Health, and Safety on a North Sea Oil Platform* (pp. 142–74). Oslo: Universitetsforlaget.

Neal, A., Griffin, M. A., & Hart, P. M. (2000). The impact of organizational climate on safety climate and individual behavior. *Safety Science,* 34(2000), 99–109.

Pomeroy, R. V., & Earthy, J. V. (2017). Merchant shipping's reliance on learning from incidents: A habit that needs to change for a challenging future. *Safety Science,* 99: 45–57.

Reason, J. (1990). *Human Error.* Cambridge: Cambridge University Press.

Reason, J. (2008). *The Human Contribution: Unsafe Acts, Accidents and Heroic Recoveries.* Manchester: Ashgate.

Roberts, S. E., & Hansen, H. L. (2002). An analysis of the causes of mortality among seafarers in the British merchant fleet (1986–1995) and recommendations for their reduction. *Occupational Medicine,* 52, 195–202.

Roberts, S. E., & Marlow, P. B. (2005). Traumatic work related mortality among seafarers employed in British merchant shipping, 1976–2002. *Occupational and Environmental Medicine,* 62, 172–80.

Roberts, S. E., Nielsen, D., Kotłowski, A., & Jaremin, B. (2014). Fatal accidents and injuries among merchant seafarers worldwide. *Occupational Medicine,* 64(2014), 259–66.

Rundmo, T. (1999). Perceived risk, health and consumer behavior. *Journal of Risk Research,* 2(3), 187–200.

Rundmo, T., Nordfjærn, T., Hestad Iversen, H., Oltedal, S., & Jørgensen, S. H. (2011). The role of risk perception and other risk-related judgements in transportation mode use. *Safety Science,* 49(2011), 226–35.

Short, J. F. (1984). The social fabric at risk: Towards the social transformation of risk analysis. Americal Sociology Association, presidential address.

Slovic, P., Fischhoff, B., & Lichtenstein, S. (1981). Perceived risk: Psychological factors and social implications. *Proceedings of the Royal Societey of London,* A376(1981), 17–34.

Weinstein, N. D. (1980). Unrealistic optimism about future life events. *Journal of Personality and Social Psychology,* 39(5), 806–20.

Yule, S. (2003). Senior management influence on safety performance in the UK and US energy sectors. PhD thesis, University of Aberdeen, Scotland.

Zohar, D. (1980). Safety climate in industrial organizations: Theoretical and applied implications. *Journal of Applied Psychology,* 65(1), 96–102.

Zohar, D. (2010). Thirty years of safety climate research: Reflections and future directions. *Accident Analysis and Prevention,* 42(5), 1522–77.

7 Design for safety

Margareta Lützhöft and Viet Dung Vu

Introduction

Design plays an important role in the maritime industry and is vital to the construction of ships, maritime equipment, management systems, as well as rules and regulations. Back in the great days of sail, vessels were designed and built on the basis of practical experience; ship construction was predominantly a skill. The advance of science and technology changed ship design into a complicated combination of art and science, delivering a dramatic increase in the size, speed and technical complexity of ships.

Throughout the history of shipbuilding, naval architects have looked at ships mainly from technical and economical points of view. Generations of ship designers have done a great job in making seagoing vessels more seaworthy, faster and with an increased transport capacity and lower fuel consumption. Great effort has also been made in improving maritime safety. Seafarers are now assisted by highly advanced automated systems and sensors, which logically should allow for safer shipboard operations and navigation. However, the high numbers of accidents and casualties at sea, with 3,296 cases in 2016 alone (European Maritime Safety Agency (EMSA), 2016), indicate a different picture.

Reports suggest that human errors are blamed for most maritime accidents, which raises the question: why do humans still make errors despite countless improvements in design and technology to assist them?

The design of technology influences the way people work and can have a big impact on how people perform. There are many cases where design issues actually bring negative effects to marine operations, as seen in the grounding of the *CFL Performer* (2008) and the *Ovit* (2013) as well as many other accidents. In the case of the *CFL Performer*, the complexity of the operation of the ECDIS (Electronic Chart Display and Information System) with regard to the number of user settings required led to the mistake of the officer in not setting up a watch vector. Consequently, although the safety contour was set at 30m, its associated alarm did not activate when the vessel crossed into shallow water (Marine Accident Investigation Branch (MAIB), 2008). Similarly, the MAIB also found 'several of the features of the … ECDIS on board Ovit were … difficult to use',

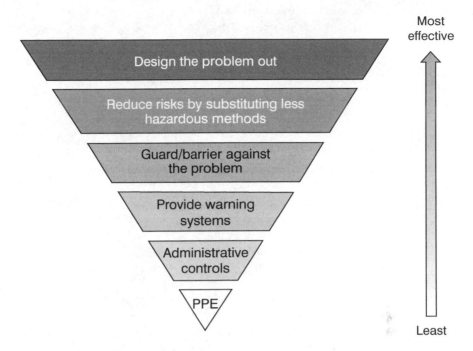

Most
effective

Least

Figure 7.1 Hierarchy of HF issues control measures

(Note: PPE = Personal Protective Equipment)

which rendered the system ineffective and contributed to the grounding (MAIB, 2014). These accidents raise awareness on the issues in the interaction between individuals, technology, systems and organizations in maritime systems, which are also referred to as Human Factors (HF) issues.

Following the hierarchy of hazard controls, HF issues can be addressed using the control measures shown in Figure 7.1.

Warning the operators of the issues in the systems is required under the provision of the International Safety Management (ISM) Code but is the least effective control measure on the scale. The issues remain in the system, and, due to the mental and physical demands of their work, crew members cannot be continually aware of the existence of a problem.

Training programmes such as Bridge Resource Management (BRM), ECDIS and type-specific training courses aim to implement best practice among crew members. However, similar to warning the crew of the problems, training does not completely eliminate the problems as we are merely adding to the workload of the crew.

A more effective solution would be installing barriers to isolate the hazards. With this approach, the crew are protected from being accidentally exposed to hazards. However, these barriers can be bypassed, removed or subject to failures.

The ultimate control measure is eliminating or removing the risk. Designing systems without HF issues is not only the most effective risk control measure but also required by maritime rules and regulation, as seen in the SOLAS (Safety Of Life At Sea) regulation V/15 regarding the design of ship bridges, bridge equipment and procedures:

> All decisions which are made for the purpose of applying the requirements of regulations 19, 22, 24, 25, 27 and 28 and which affect bridge design, the design and arrangement of navigational systems and equipment on the bridge and bridge procedures shall be taken with the aim of:
> 1.1 facilitating the tasks to be performed by the bridge team and the pilot in making full appraisal of the situation and in navigating the ship safely under all operational conditions;
> 1.2 promoting effective and safe bridge resource management;
> 1.3 enabling the bridge team and the pilot to have convenient and continuous access to essential information which is presented in a clear and unambiguous manner, using standardized symbols and coding systems for controls and displays;
> 1.4 indicating the operational status of automated functions and integrated components, systems and/or sub-systems;
> 1.5 allowing for expeditious, continuous and effective information processing and decision-making by the bridge team and the pilot;
> 1.6 preventing or minimizing excessive or unnecessary work and any conditions or distractions on the bridge which may cause fatigue or interfere with the vigilance of the bridge team and the pilot; and
> 1.7 minimizing the risk of human error and detecting such error if it occurs, through monitoring and alarm systems, in time for the bridge team and the pilot to take appropriate action.

The influence of HF design on safety has been acknowledged (Earthy & Sherwood Jones, 2011), and relevant parties have started taking steps in developing a new approach to design systems free from the HF issue, or in simpler words, to 'design the problem out'. We acknowledge that the SOLAs regulation (V/15) is difficult to quantify and design for in the traditional engineering fashion. This chapter will discuss the importance of design in relation to maritime safety and explore methods that can be used to help achieve safer life and work at sea through improved designs and design processes.

Background

Design is everywhere. Take a look around you: the outfit you are wearing is designed, the pen you write with every day is designed, the computer software you use daily is designed, this book chapter was also designed; design is everywhere. This phenomenon explains why there are countless definitions among both the research and design communities and why it is so hard to clearly define what 'design' is.

Ralph and Wand (2009, p. 6) introduce a formal definition of design as 'a specification of an object, manifested by an agent, intended to accomplish goals, in a particular environment, using a set of primitive components, satisfying a set of requirements, subject to constraints'. This definition is neutral, and when applied to the ship design context, we can establish the appropriate understanding of its particular process.

Within the maritime domain, the design objects are variable. They can be the ships, shipboard equipment or port facilities; they can also be management systems, regulations, rules, guidelines and procedures. Let us take a cargo ship as an example. A modern ship is an extremely sophisticated piece of engineering, due to both its size and functional complexity. Typically, a ship consists of several compartments, the largest of which are the cargo holds. Holds can have different forms, built and coated with various materials and equipped with several devices and sensors depending on the type of cargo the ship carries. The machinery space of the ship is the location of the main and auxiliary engines, generators, pumps, steering gears and other mechanical equipment. The crew inhabit the accommodation area, and the navigation bridge is located on top of the superstructure. Ships are built at the shipyards, and because of their size and technical difficulty, they have to be carefully designed before the building starts – a work which takes months or even years to complete.

The design of a ship is her specification, created by naval architects in the form of technical drawings. These drawings are the model of the ship, including her exterior as well as interior structure together with locations and details of compartments and equipment. Using Ralph and Wand (2009) as a guide, such specifications must ensure that the future ship can safely and economically transport cargo (*goals*) under the specific maintenance state and the conditions of the trading area (*environment*). At the same time, the naval architects must also ensure their design can satisfy the ship owner's requirements such as the type of cargo, cargo capacity, desired speed or area of operation while complying with relevant rules and regulations (*constraints*), e.g. for environment and safety.

The shipyard carries out the construction of the ship based on the design, and the design will remain the guiding star throughout the process of building. Should the design be modified, so is the future ship. This principle applies to all design objects, whether it is a ship, a radar system, a training programme or a set of regulations. The specification should determine the characteristics of the design object. Thus, a good design will result in a product which smoothly helps us perform our duties while a bad design will make work harder to perform efficiently. However, even if your equipment is designed well, the installation onboard can have an influence. Outfitting and installation is done by subcontractors working to relatively general requirements. They have to make on the spot 'design' decisions which can depend on something as mundane as the length of cable they brought on board. Thus, the influence on ergonomics of arrangements should not be underestimated. This chapter will focus mainly on the design of the hardware in shipping – ships and maritime technologies.

Design can enhance safety

This section begins with a discussion of the impact of design on safety and follows with the introduction of factors to consider promoting safety by design and examples of their application in the industry.

The impact of design on safety

Safety should be one of the primary concerns of the maritime community, due to the hazardous and unpredictable nature of the maritime working environment. We can assume that this is a concern for seafarers, but it is not clear that it is so for owners – unless we also include the ship and cargo. Through the years, non-governmental organizations and state authorities have made efforts to improve safety in the industry. The development of science has made technology a key instrument in making the sea a safer place. Despite the fact that all technology designs are safety-prioritized and many designs are even created with the sole purpose of improving safety, accidents still happen. The most common explanation for accidents and casualties at sea is human error. Setting aside factors beyond human control, several researchers have suggested that human error is the most contributing factor in maritime accidents (Dhillon, 2007; Macrae, 2009; Rothblum, 2000). As a result, safety would be significantly improved if human erroneous action could be avoided.

The maritime system is a people system. People interact with technology, the environment and organizational factors (Grech, Horberry, & Koester, 2008; Rothblum, 2000). Safety is decided by the way humans interact with other system components and, most of the time, this has been predetermined in designs. The human as a living component is expected to provide necessary adaptation and interact with other components in the requested manner. Traditionally, this has been facilitated through training, familiarization and following procedures. The marine radar is one example. It was designed to assist human vision in poor visibility and, ultimately, improve safety at sea. The seafarers are trained to operate marine radar and must follow proper practice when working with them. When the human fails to comply with the procedure and carries out erroneous actions, the faulty interaction breaks the link between elements and the cooperation breaks down. This is a common scenario of accidents in the maritime domain, and whenever this happens, the human is blamed.

Demonstration of such deviation from standard practice can be observed in the case of the *Seastreak Wall Street* allision. On 9 January 2013, while approaching berth at lower Manhattan, New York, the high-speed ferry *Seastreak Wall Street* struck a pier. The investigation found that the captain had forgotten to return the propulsion system to Combinator mode after switching to Backup mode earlier in the voyage. Consequently, when the captain moved the engine order levers astern to reduce speed, his action actually resulted in forward acceleration. This allision is a typical example of human erroneous action. The captain should have been aware of the active propulsion control mode. He should have carried out the correct action, which would be slowing down the ship instead of accelerating. His actions resulted in injuries to 80 people on board and damage to both the ship and the pier.

However, the investigation did not end there and the mistakes of the captain revealed much more complicated issues in the man–machine cooperation on board the *Seastreak Wall Street*. The report also points out that the propulsion system onboard the ship used poorly designed visual and audible cues and failed to provide the captain with critical information about mode and control transfer status. Such faulty design can lead to mistakes in operator performance, especially in intense traffic conditions as in this case (National Transportation Safety Board, 2014).

All human beings have capabilities and limits. For instance: humans are not very good at multi-tasking. The ability to stay focused is largely affected by the surrounding environment, and our physique determines reach, strength, agility and resistance to fatigue. Time and training can help improve skills and competencies, but only to a certain level and there will always be limits to our physical, sensory and mental abilities.

Many maritime technologies designs overlook this matter because they are designed by people with a purely technical mindset who were trained to deal with problems from technical perspectives. Consequently, many designs are logically functional but lack usability. The operators are left struggling to fill the gap between their duties' requirements and the predetermined interactions (Lützhöft, 2004). Occasionally, this additional work exceeds human abilities and the human component fails, leaving the system exposed to hazards. The underlying cause, however, is not human error but rather the designs which are incompatible with human capabilities, leading to a system failure.

In fact, design aspects can have a significant impact on fatalities at sea. The main contributors are well known in shipping: slips/trips/falls, mooring and enclosed spaces. A sample of 66 deaths in British shipping 2003–12 (Roberts, Nielsen, Kotłowski, & Jaremin, 2014) was analysed from an ergonomics perspective and found that 64 per cent of the cases (42 persons) could have been mitigated by design (Sherwood Jones, 2016).

To sum up, in order to maintain safety, minimize environmental impact and achieve expected performance it is vital to maintain proper interaction between the human and other components of maritime systems. A faulty design can make such interaction complicated and occasionally impossible for the human to uphold proper interaction, resulting in human errors. Thus, it is faulty design, not 'human error', that is the primary, or latent, reason behind accidents in the maritime industry.

Human factors in design

As discussed in the previous section, bad design may increase the risk of accidents. This fact was strongly emphasized by Miller (2000, p. 7):

> You cannot overcome human errors induced by poor Design of the workplace with more training, more Manuals or written procedures, exhortations to work more safely, or threats of punitive actions for job accidents.

Hence, the best safety solution lies with the design itself, so the underlying question is: how should the ship and shipboard equipment be designed to improve safety?

As the human is the only living, adapting component in a ship system, it is important that human–system issues are carefully considered by designers to produce a design which can accommodate the needs, capabilities and limitations of humans. The 'scientific discipline concerned with the understanding of interactions among humans and other elements of a system, and the profession that applies theory, principles, data, and other methods to design in order to optimize human well-being and overall system performance' is defined as 'human factors' or 'ergonomics' (International Ergonomics Association, 2012). A design taking into account human factors will facilitate the interaction between humans and other system components, significantly reduce the probability of erroneous actions, and thus improve safety (Abeysiriwardhane, Lützhöft, & Enshaei, 2014).

Within the specific scope of the maritime industry, the International Maritime Organization (IMO) adopted the term 'human element' to describe the complex multi-dimensional issue that 'involves the entire spectrum of human activities performed by ship's crews, shore-based management, regulatory bodies, recognized organization, shipyards, legislators, and other relevant parties' (International Maritime Organization [IMO], 2003, p. 3).

Despite the usage of different wording, all three terms – ergonomics, human factors and human element – mean the same approach in designing with due consideration to human needs and abilities to improve effectiveness and efficiency and improve overall system performance. For this reason, these three terms will be used interchangeably.

There are many methods of addressing the human element in the maritime domain. Within this chapter, we will use the framework of addressing the human element, which was introduced by Earthy, Sherwood Jones, and Squire (2016) in the Human Element Alert! Project bulletin issues 11 and 40: see Figure 7.2. Following the framework, the human element can be addressed through human resources, social and organizational and human factors considerations. The first two groups can and should be addressed at management and operational levels while human factors considerations can generally be addressed in system/vessel design phase. These considerations are connected; the more effort put into the design phase, the less should be needed in operations.

The framework divides human factors into nine aspects: Habitability, Maintainability, Security, Occupational Health and Safety, Manoeuvrability, Controllability, Survivability, System safety, and Workability. Each of these aspects will be explained below with reference to the introduction by Myles (2015).

Habitability

Habitability represents the quality of a vessel that allows people to live and work in a safe and productive manner. On board ships, habitability is expressed by the acceptability in terms of whole-body vibration, noise, indoor climate, lighting,

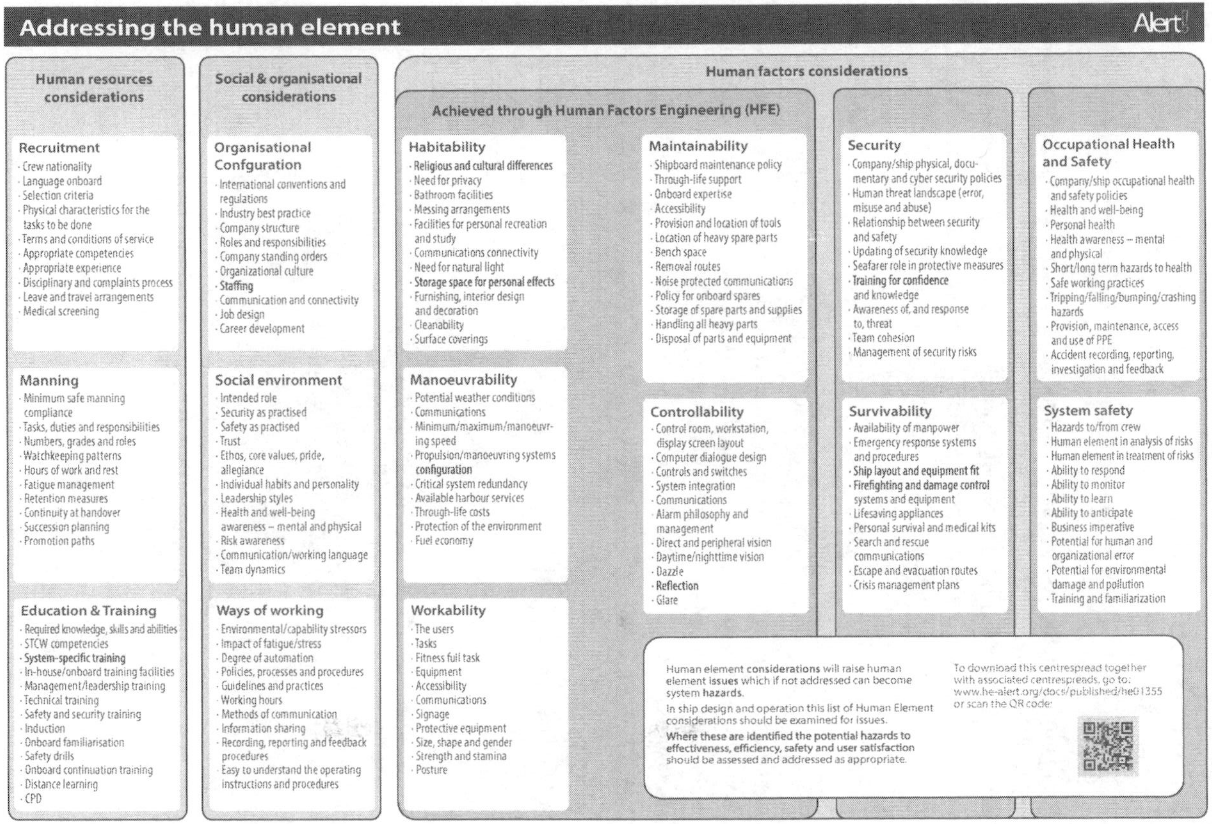

Addressing the human element

Human resources considerations

Recruitment
- Crew nationality
- Language onboard
- Selection criteria
- Physical characteristics for the tasks to be done
- Terms and conditions of service
- Appropriate competencies
- Appropriate experience
- Disciplinary and complaints process
- Leave and travel arrangements
- Medical screening

Manning
- Minimum safe manning compliance
- Tasks, duties and responsibilities
- Numbers, grades and roles
- Watchkeeping patterns
- Hours of work and rest
- Fatigue management
- Retention measures
- Continuity at handover
- Succession planning
- Promotion paths

Education & Training
- Required knowledge, skills and abilities
- STCW competencies
- **System-specific training**
- In-house/onboard training facilities
- Management/leadership training
- Technical training
- Safety and security training
- Induction
- Onboard familiarisation
- Safety drills
- Onboard continuation training
- Distance learning
- CPD

Social & organisational considerations

Organisational Configuration
- International conventions and regulations
- Industry best practice
- Company structure
- Roles and responsibilities
- Company standing orders
- Organizational culture
- **Staffing**
- Communication and connectivity
- Job design
- Career development

Social environment
- Intended role
- Security as practised
- Safety as practised
- Trust
- Ethos, core values, pride, allegiance
- Individual habits and personality
- Leadership styles
- Health and well-being
- Risk awareness — mental and physical awareness
- Communication/working language
- Team dynamics

Ways of working
- Environmental/capability stressors
- Impact of fatigue/stress
- Degree of automation
- Policies, processes and procedures
- Guidelines and practices
- Working hours
- Methods of communication
- Information sharing
- Recording, reporting and feedback procedures
- Easy to understand the operating instructions and procedures

Human factors considerations

Achieved through Human Factors Engineering (HFE)

Habitability
- **Religious and cultural differences**
- Need for privacy
- Bathroom facilities
- Messing arrangements
- Facilities for personal recreation and study
- Communications connectivity
- Need for natural light
- **Storage space for personal effects**
- Furnishing, interior design and decoration
- Cleanability
- Surface coverings

Manoeuvrability
- Potential weather conditions
- Communications
- Minimum/maximum/manoeuvering speed
- Propulsion/manoeuvering systems **configuration**
- Critical system redundancy
- Available harbour services
- Through-life costs
- Protection of the environment
- Fuel economy

Workability
- The users
- Tasks
- Fitness full task
- Equipment
- Accessibility
- Communications
- Signage
- Protective equipment
- Size, shape and gender
- Strength and stamina
- Posture

Maintainability
- Shipboard maintenance policy
- Through-life support
- Onboard expertise
- Accessibility
- Provision and location of tools
- Location of heavy spare parts
- Bench space
- Removal routes
- Noise protected communications
- Policy for onboard spares
- Storage of spare parts and supplies
- Handling all heavy parts
- Disposal of parts and equipment

Controllability
- Control room, workstation, display screen layout
- Computer dialogue design
- Controls and switches
- System integration
- Communications
- Alarm philosophy and management
- Direct and peripheral vision
- Daytime/nighttime vision
- Dazzle
- **Reflection**
- Glare

Security
- Company/ship physical, documentary and cyber security policies
- Human threat landscape (error, misuse and abuse)
- Relationship between security and safety
- Updating of security knowledge
- Seafarer role in protective measures
- **Training for confidence** and knowledge
- Awareness of, and response to, threat
- Team cohesion
- Management of security risks

Survivability
- Availability of manpower
- Emergency response systems and procedures
- **Ship layout and equipment fit**
- **Firefighting and damage control** systems and equipment
- Lifesaving appliances
- Personal survival and medical kits
- Search and rescue communications
- Escape and evacuation routes
- Crisis management plans

Occupational Health and Safety
- Company/ship occupational health and safety policies
- Health and well-being
- Personal health
- Health awareness — mental and physical
- Short/long term hazards to health
- Safe working practices
- Tripping/falling/bumping/crashing hazards
- Provision, maintenance, access and use of HFE
- Accident recording, reporting, investigation and feedback

System safety
- Hazards to/from crew
- Human element in analysis of risks
- Human element in treatment of risks
- Ability to respond
- Ability to monitor
- Ability to learn
- Ability to anticipate
- Business imperative
- Potential for human and organizational error
- Potential for environmental damage and pollution
- Training and familiarization

Human element considerations will raise human element **issues** which if not addressed can become system hazards.

In ship design and operation this list of Human Element considerations should be examined for issues.

Where these are identified the potential hazards to effectiveness, efficiency, safety and user satisfaction should be assessed and addressed as appropriate.

To download this centrespread together with associated centresgeouts, go to: www.he-alert.org/docs/published/he01355 or scan the QR code

Figure 7.2 Addressing the human element

and physical and spatial characteristics, according to prevailing research and standards (American Bureau of Shipping (ABS), 2016), the most notable of which is the Maritime Labour Convention (International Labour Organization, 2006).

The design of the vessel must be able to provide comfortable, clean (cleanable) and convivial accommodation, washing and toilet facilities, mess rooms, group meeting and exercise areas and recreational spaces with due consideration to the crew's variation in anthropometric characteristics, genders and culture diversity (Alert! Project, 2006).

Poor lighting, whether too bright or too dim, can have an adverse effect on human performance and physiological as well as psychological well-being. Similarly, whole-body vibration can lead to discomfort, degraded performance, and in some cases can even lead to chronic health problems such as back pain, musculoskeletal disorders and temporary physiological changes (Grech et al., 2008).

Inappropriate levels of noise can decrease vigilance during watchkeeping duties, affect sleep and rest quality and consequently lead to fatigue and degraded performance. Noise and vibration require special attention since routine shipboard operations involve working around noisy machinery, and therefore it is essential that measures are taken to minimize the effects of noise, such as insulation, noise absorbent flooring, large diameter propeller, shock absorbers engine and rubber suspension exhaust system (Alert! Project, 2014).

Furthermore, the design of working and accommodation areas must also consider the impact of climatic conditions on human performance and health, particularly the temperature, humidity and the quality and circulation of air. The purpose is to provide an environment which is suitable to promote optimal task performance and crew physiological well-being.

Maintainability

From the engineering point of view, maintainability is defined to be the probability that a failed component or system will be restored to the condition of operational effectiveness within a period of time when maintenance is performed following prescribed procedures (Ebeling, 2004; Smith, 2011). Thus, maintenance is essential for the safe and efficient operation of all systems.

From the operators' perspective, however, maintenance is a complex task often involving the removal and replacement of several components, which requires high vigilance and skills and is commonly performed in difficult working conditions under time pressure. As a result, maintenance work is especially vulnerable to error (Reason & Hobbs, 2003).

Design plays an important role in deciding the outcome of maintenance tasks. Bad ergonomic features such as lack of access, restricted space to manoeuvre, or components that can be incorrectly fitted will hinder the ability to perform effective maintenance duties and under certain conditions can even trigger critical events, as seen in the accident that led to the death of a crew member while carrying out routine lifeboat maintenance onboard the passenger ship *Volendam* in Lyttelton, New Zealand (Transport Accident Investigation Commission, 2011).

To promote the conduct of maintenance duties in safe and effective manners, it is essential that the design of the ship and shipboard equipment provide design solutions to allow for operational maintenance tasks to be rapid, safe and effective. Such consideration should include habitability factors as well as the nature of the tasks, equipment access and the requirements and capabilities of the crew members performing the tasks.

Manoeuvrability

Statistics of accident data suggest that most collision and grounding accidents originate from actions performed on the navigation bridge (ABS, 2006). The success of collision avoidance and safe navigation of ships largely depends on the safe handling of ships, which requires the effective interaction between the human operators – the seafarers – and the manoeuvring characteristics of the ship. While every ship regardless of her manoeuvrability can be handled, some ships with poor manoeuvrability can be very difficult to handle under certain circumstances such as high traffic density or restricted area to manoeuvre and can pose a threat to navigation safety.

Therefore, for the safety of ships, people onboard and the marine environment, vessels must be designed and constructed in a way to have the most appropriate manoeuvring capabilities consistent with the intended role, manning and operating pattern of the ship, taking into account propulsion and manoeuvring configuration, communication, critical system redundancy, weather conditions and harbour services, with the aims of cost efficiency, fuel economy and environmental protection (Alert! Project, 2016).

Controllability

A consideration of the integration of a working system which takes into account human capabilities and limitations as well as technologies can help achieve efficient control of systems and ship. This includes knowledge of human physiology and psychology but also design spanning from layout to dialogue design levels. The following points are included in controllability:

- Control room, workstation, display screen layout
- Computer dialogue design
- Controls and switches
- System integration
- Communications
- Alarm philosophy and management
- Direct and peripheral vision, daytime/night-time vision, dazzle, reflection and glare.

The issue of design for controllability has become more important because of increased complexity of marine equipment, particularly with the application of automation and computer-based systems. Resources for achieving this are mainly found in the ISO 9241-series of standards on the ergonomics of human–system interaction and in guidelines from classification societies, for example, the ABS (Australian Bureau of Shipping) 86 Guidance Notes on the applications of ergonomics to marine systems.

Workability

Workability represents the capacity of the vessel and her equipment suitable for the intended work situation. Due consideration should be given to the users, tasks, equipment, materials and procedures and the physical and social aspects of the working environment (Lloyd's Register, 2008), including:

- The users, their tasks and their fitness for task
- Equipment and accessibility
- Communications and signage
- Protective equipment
- Size, shape and gender, strength and stamina, and posture.

In order to make something for someone, you need to know who they are and what they are trying to do. In one Japanese management theory, this is addressed by 'going to the gemba'. This means finding out what is going on by going 'to the floor'. Consult users about their jobs, find out about their capabilities, and let them show you how they work. Do not ask them what they want to have; users are not designers – it is their professional skills we want to support, not their ship design skills.

Security

This consideration has evolved from mainly physical security as addressed by the ISM and ISPS (International Ship and Port Facility Security) codes to the cyber landscape. Areas include the following (Alert! Project, 2016).

- Company/ship physical, documentary and cyber security policies
- Human threat landscape (error, misuse and abuse)
- Relationship between security and safety
- Updating of security knowledge
- Seafarer role in protective measures
- Training for confidence and knowledge
- Awareness of, and response to, threat
- Team cohesion
- Management of security risks.

A number of resources are available, mainly in the form of training, awareness and policy. The central aim is naturally to protect the seafarer and the ship, but also the business and the environment. Not many of these issues are addressed in design, but we would suggest that seafarers can be protected to some degree by physical design and, not least, by having a plan for addressing cyber security in the design phase.

Survivability

Survivability is not just about the adequacy of firefighting, damage control, lifesaving and security facilities; it is also about having the correct resources, training and procedures in place to ensure the safety of the ship and to protect the health, safety and well-being of its seafarers (Alert! Project, 2015a). Survivability includes:

- Availability of manpower
- Emergency response systems and procedures
- Ship layout and equipment fit
- Firefighting and damage control systems and equipment
- Lifesaving appliances
- Personal survival and medical kits
- Search and rescue communications
- Escape and evacuation routes
- Crisis management plans.

This area is critical, and one where many unfortunate accidents happen to crew trying to manage an emergency. An unacceptable number of fatalities result from drills, especially with lifeboats and rescue boats. Between 1989 and 1999 lifeboats and their launching systems have cost the lives of 12 professional seafarers or 16 per cent of the total lives lost on merchant ships. Additionally, 87 seafarers were injured (MAIB, 2001). A notable accident occurred in 2014 on board the Maltese-flagged bulk carrier *Aquarosa* while the ship was in the Indian Ocean on her voyage from Singapore to Fremantle, Western Australia. The ship's freefall lifeboat was inadvertently released during a routine inspection, causing serious injuries to the second engineer, who was the only person in the lifeboat at the time. Subsequent investigation by the Australian Transport Safety Bureau (ATSB) found the design of the lifeboat's on-load release system to be a contributing factor to the accident (Australian Transport Safety Bureau, 2015). In addition, statistics show lifesaving appliances to be one of the ten most focused-on areas of safety recommendations issued by maritime investigative bodies of EU States between 2011 and 2015 (EMSA, 2016), making survivability a crucial factor to consider in addressing the human element in the maritime industry.

Occupational health and safety

Designers must take into account the effect of work, the working environment and living conditions on the health, safety and well-being of the person. The

MLC (Maritime Labour Convention) emphasizes the rights of every seafarer to a safe and secure workplace that complies with safety standards; to fair terms of employment; to decent working and living conditions on board ship; and to health protection, medical care, welfare measures and other forms of social protection. The considerations include:

- Company/ship occupational health and safety policies
- Health and well-being
- Personal health
- Health awareness – mental and physical
- Short/long-term hazards to health
- Safe working practices
- Tripping/falling/bumping/crushing hazards
- Provision, maintenance, access, and use of personal protective equipment (PPE)
- Accident recording, reporting, investigation and feedback.

Further support is available in the ILO (International Labour Organization) Guidelines for implementing the occupational safety and health provisions of MLC 2006 (Alert! Project, 2015b). Additionally, there are a number of ILO codes on worker protection and health and an IMO guideline on an occupational health and safety (OHS) programme.

System safety

This consideration is about the effect of people and their behaviour on the safety of systems. It is the interface between risk management and human factors and concerns the effect human behaviour may have on the safety of systems. It is important to include a judgement of risk when planning the design and taking into account hazards in underlying task descriptions and analyses. In this consideration, it is prudent to perform human reliability analyses and to consider resilience of the system. Resilience is briefly described as supporting safe operation as opposed to minimizing risk, by ensuring that an organization can continue operating in the face of threats. An organization's resilience can be operationalized by auditing and strengthening the ability to respond, monitor, learn and anticipate – which can all be supported by good user-centred design.

- Hazards to/from crew
- Human element in analysis of risks
- Human element in treatment of risks
- Ability to respond, monitor, learn and anticipate
- Business imperative
- Potential for human and organizational error
- Potential for environmental damage and pollution
- Training and familiarization.

Although not all the human factors aspects in the framework are equally applicable to design and sometimes overlap, their implementation will nevertheless help create better workplaces with enhanced usability. The following section will discuss notable success stories of ship design cases to explain further the positive influence human factors inclusion may have.

Notable ship design cases

The framework of addressing the human element can be applied to all design projects. The actual application, however, varies among different cases. This section considers the implementation of human factors dimensions in three design projects, namely the development of the Tamar-class lifeboat, operated by the Royal National Lifeboat Institution (RNLI); the Service Operations Vessel 9020, designed by Damen Shipyards Group; and the Pure Car/Truck Carrier (PCTC) *Harvest Leader*, designed by Andreas Shipping Ltd.

The Tamar lifeboat class

The Tamar lifeboat design (see Figure 7.3) is an early initiative to involve users and design a vessel to their needs. The main motivations were championing organizational support and a higher duty of care than the merchant sector. The crew members are volunteers and therefore have greater protection under law, and a safe and comfortable work environment is also needed in order to retain manpower supply. Lifeboats are often called out in adverse weather conditions and under time-critical circumstances. Furthermore, their size and shape make them subject to severe effects of heavy pitching and rolling which can significantly hinder the crew's performance and bring risks and potential health issues. Additionally, the crew consists of up to 90 per cent volunteers with little or no maritime professional background. Therefore, to promote the success of search and rescue operations as well as to protect the well-being of people on board, it is required of the designers to ensure the technology on board is safe, effective and intuitive for these 'ordinary people' to use (Chaplin & Nurser, 2007).

The decision was to bring the boats to the crew members by applying ergonomics, and the process started with a risk assessment based on operational feedback and a task hazard analysis (Chaplin & Nurser, 2007). It was found that the primary concern in the lifeboat's design was the potential of lumbar spinal injury caused to the operator by the lifeboat's motion under heavy sea conditions – so the seat became a major focus. The design team employed computer-modelling techniques to model boat and wave motion and the motion of the human lumbar spine to identify different types of injury sustained on different sea states and seat design. The results of this research led to new seat design for comfort, shock absorption and to improve the ability to control the vessel.

Given the stationary position of the operators, the Tamar lifeboat is fitted with an integrated electronic Systems and Information Management System (SIMS), appropriately described as 'Safety In My Seat' by lifeboat crew. The functions

Figure 7.3 A Tamar-class lifeboat of the RNLI fleet

include: the navigation of the lifeboat, including direction finding, radar and charting; radio communications and CCTV; and the mechanics of the lifeboat including the engines, bilge and electrics (Royal National Lifeboat Institution, 2014). The integration of functions and data greatly reduce the need for the operator to move around the boat at sea. In addition, this system allows for flexibility in task sharing which makes it possible to have a crew of mixed experience and expertise.

RNLI staff and crew are involved in all stages during the development of the Tamar class. During the operation of the vessel, crew feedback is continuously collected to generate data for future improvements. A number of examples of how human factors dimensions were implemented are shown in Table 7.1.

As said by Neil Chaplin, the RNLI's principal naval architect: 'Seven years is indeed a long time to develop one boat; however, with the seats and SIMS we now have components that can be applied to future lifeboat designs and perhaps retrofitted to current ones to enhance the safety of our volunteer crews' (Alert! Project, 2006, p. 7). The RNLI Tamar boats are an example of augmenting human capability with usable technology that combines with their values and training to achieve the very best solution for people in danger.

Damen Service Operations Vessel 9020

The Damen Service Operations Vessel 9020 (see Figure 7.4) is an example of how human element issues can be addressed in operational design. The ship was built for the deployment and retention of offshore support and maintenance engineers. This design used a new operational concept which needed iterative involvement of stakeholders. The development of this vessel has been driven by functional equipment, based on feedback from potential end-users. Human element issues have been included in all levels of design, from hull form through the general arrangement and the selection and design of systems, to detailed ergonomics. How this operational design addresses human element issues is shown in Table 7.2.

Table 7.1 Results of implementing human factors dimensions in the design of the Tamar lifeboat class

Aspect	Comment
Design process	User and stakeholder involvement through iteration, from concept to operation. With station trials across the country to take any ideas in for the next Tamar (continual improvement).
Seakeeping ability	The Tamar class is self-righting, returning to an upright position in the event of capsize.
Workability	SIMS supports task sharing among the crew and saves space by replacing separate systems and equipment. Space layout includes PPE, grab rails, slip-resistant coatings, signage, etc. Guard rails, two gate sections on each side to allow different boat/pontoon height. Tamar carries a daughter boat for accessing difficult areas. To deploy the daughter boat there is a floodable recess which also allows casualties to be easily handled on board.
Controllability	The seats incorporate controls such as throttles, joystick and a trackball to operate SIMS. Communication through headsets. If all back-ups fail, the lifeboat can still be operated using conventional methods. Two control stations are used for different phases of mission. At the upper station there are two throttle controls to allow the helm to gain the best view of the side in use.
Manoeuvrability	The Tamar lifeboat is very agile and has a top speed of 25 knots. She has twin engines and a bow thruster. Out of sea manoeuvrability is improved by multiple launch and recovery options. Mast section on hydraulics to allow easy and quick movement between stored and active states.
Maintainability	The Tamar's propellers and rudders are protected by tunnels and the keels are steel lined. Engine room floor layout allows complete access to the whole engine for easy maintenance. Labelling and easy access to all serviceable components and sampling points.
Habitability	The lifeboat can carry 118 survivors standing. Special crew seats for comfort, shock absorption and control. Layout optimized for safe and efficient teamwork during rescue missions.
System safety	Interface designed for situation awareness, shared task load, safety through remote monitoring and control, redundancy and simplified training. The coxswain has control, thus allowing to give control to who needs it, so no mixed messages.
Occupational health and safety	The shock-absorbing seats enhance crew comfort and safety. There is belted seating for 10 survivors. Grab rails, layout, best possible protective and survival equipment are provided in combination with very high levels of training and drilling.
Security	SIMS uses an Ethernet local area network architecture for fast and reliable communication between workstations. If a single point failure occurs, system data integrity is not affected.
Survivability	Design loads are set to higher than expected use, and a safety factor is used, meaning that even if the lifeboat is taken out beyond expected performance it will withstand the loads. Secondary buoyance is built into hull. Additionally, she carries a salvage pump. The control system is made with rugged and salt water-resistant hardware and has no moving parts, which makes it robust and shock-resistant. The architecture is resilient as back-up systems reduce the risk of complete system failure.

Table 7.2 Results of implementing human factors dimensions in the design of the Damen vessel 9020

Aspect	Comment
Design process	User and stakeholder involvement, iteration.
Seakeeping ability	The hull form is longer (90m) and shallower (4.6m) than a comparable conventional PSV (Platform Supply Vessel). The bow section has been lowered by 1.5m compared to conventional platform supply vessels, which significantly reduces slamming in heavy seas.
Workability	New context of use for operation further offshore where crew transfer within a day becomes impractical; voyages may last a month. 80% weather operability in wave heights of up to 3m. Power savings compared with conventional PSV. Interior zoned to assist workflow and storage. Walking distances minimized. Facility locations match activity sequences. User tasks considered for logistical flows. Air-conditioned workshop facilities to store sensitive parts and tools.
Controllability	Single deckhouse for bridge and site management.
Manoeuvrability	DP (dynamic positioning) available, but designed to be largely used non-DP.
Maintainability	Separation of clean and dirty areas.
Habitability	Longer hull, shallower draught for more comfortable motion. Worker accommodation midships to reduce pitch motion effects. Single-occupancy cabins, fitness centre and internet/movie services. Maximum use of natural light. Comfortable décor scheme.
Connectability	Offices, conference room, recreation dayrooms, reception room, hospital, drying room, all cabins provided with internet telephone and satellite TV.
System safety	For a high level of redundancy, alternative transfer methods are included such as helicopter, crane or a Damen-designed rigid hull inflatable boat (RIB).
Occupational safety and health	Crew transfer hazards reduced by motion-compensated gangway from a stable DP2 platform.

The PCTC Harvest Leader

Cargo ships account for the majority of the world fleet (Equasis, 2015). Therefore, an attempt to implement human factors and engineering in the design of merchant ships can have a powerful effect on the maritime industry. The PCTC vessel *Harvest Leader* (see Figure 7.5) designed by Andreas Shipping Ltd for NYK Line is an example of such effort (Bialystocki, 2016).

The process of designing the ship followed an unusual practice in the industry, where crew feedback was gathered and subsequently analysed by the design team alongside the owner's designers, the managers and the shipyard, in order to find a solution to apply on the design. Such practice resulted in various design improvements on the ergonomic aspects of the vessel: see Table 7.3. Personal communication with the project leader revealed that, although some of the changes were made at the later stage, thus leading to some increase of cost, this

Figure 7.4 Damen Service Operations Vessel

Figure 7.5 The *Harvest Leader* off the coast of Point Lonsdale, Australia, (photograph Lester Hunt, http://www.marinetraffic.com/en/ais/details/ships/shipid:1120219)

Table 7.3 Results of implementing human factors dimensions in the design of the PCTC *Harvest Leader*

Aspect	Comment
Design process	User and stakeholder involvement through iteration. The involvement of users was deemed unsuitable and insignificant at the early design stages but was much more fruitful at the subsequent detailed design stage.
Seakeeping ability	Due to the commercial purpose of a merchant ship, the *Harvest Leader* was designed and built to satisfy the seakeeping ability requirements of the regulations.
Workability	Vast improvements were implemented and eventually achieved positive outcomes regarding workability. Alternation of cargo hold fans location help preventing obstruction to visibility. The bridge control console was separated into wing control console and instruments console. Both were planned to facilitate ease of use and intuitive operation. The systems were installed with the principle of being as uncomplicated as possible, notably the removal of unused equipment usually provided by the shipyard, the design of modules of 'plug and play' type allows for easy and efficient maintenance.
Controllability	Cameras were installed at critical locations. Additional telephones and PA speakers were provided in working and accommodation areas to improve internal communication. The integration of all alarms from bridge systems into one Bridge Alarm Monitoring Systems and the use of a user-friendly Integrated Monitoring, Alarm, and Control Systems allows for easy monitoring and control of bridge, engine and machinery systems. The equipment is standardized within the owner's fleet to support familiarization and adaptation.
Manoeuvrability	The *Harvest Leader* was designed to achieve service speed of about 20.0 knots and equipped with bow and stern thrusters to improve manoeuvrability.
Maintainability	All machinery in the engine room was arranged with 360° access, and less frequently attended spaces such as cargo hold ventilation trunks were also provided with safe and comfortable means of access to allow for easy maintenance. Workshops in the engine room were designed and equipped with additional equipment such as ultrasonic cleaner and lifting devices to further support maintenance.
Habitability	Five extra cabins were provided to accommodate irregular workers. The illumination level was raised high above the flag requirements for crew comfort, and in working areas it increases safety through easier task performance. Moreover, several spaces without an illumination level requirements were also specified by the owner's request such as 300 lux in the wheelhouse and engine control room, 200 lux in the engine room workshop. Additional noise reduction measures were taken to maintain comfortable noise level on board. A 5-blade propeller was installed to reduce the effect of propeller on the vibration level. Crew's living and common spaces were enlarged to allow more comfortable habitation.
System safety	Bridge and engine control systems are designed to improve situation awareness, increase ease of use, and many preventative measures were applied to prevent crew mistakes.

Aspect	Comment
Occupational safety and health	Working and exposed decks are painted with anti-slip coating to avoid slipping. Kick plates are installed on each side of fixed ramps to prevent fittings from rolling and falling. Illumination, noise and vibration levels were controlled to maintain at high safety level as required by latest IMO standards. The increase of manning level helps maintain sufficient resting hours.
Survivability	Several improvements were implemented which exceed the minimum safety standards. Fire detectors were installed in all cabins and fire hydrants with hoses were provided in the accommodation area. The crew was provided with additional life jackets at the muster stations to allow for immediate access in cases of emergency. Oversize and childsize life-saving appliances were added.

was absorbed by savings in other parts of the build and subsequent expense, especially maintenance cost. During the ship's life cycle, crew feedback will continuously be collected and taken into account when considering changes in arrangements of equipment or implementation of new technologies.

How shall we design?

In the previous two sections, we discussed the influence of design on the safety level of the whole system, the necessity of addressing the human element in design, as well as certain cases of optimal design with due regard to human factors. This section will summarize the most important aspects of a good design process, which require serious consideration from ship designers and equipment manufacturers, including human-centred design (HCD), regulations, standards and guidelines, as well as challenges in the design process.

Human-centred design

As seen in the cases of the Tamar lifeboat, the Damen Service Operations Vessel 9020 (Walk2Work) wind farm support vessel and the PCTC *Harvest Leader*, the designers applied input from users to implement human factors aspects in the designs, which resulted in the improvement of effectiveness, efficiency, satisfaction and overall safety. In such cases, the designers follow a process of 'applying Human Factors and ergonomics knowledge and techniques to minimize human error, enhance effectiveness and efficiency, improve human working conditions, and counteract the possible adverse effects of use on the health, safety, and performance of the mariner', which is known as human-centred design.

The standard ISO 9241-210:2010 sets out the principles of HCD, which can be briefly summarized as follows. The future or potential users are involved through the design and development; the designers cooperate with the users to understand user needs, the nature of the tasks and the product's intended working environment. Such data are combined with human factors and usability knowledge and techniques to create highly usable systems, which can improve productivity, enhance user well-being and reduce risk of harm (International Organization for Standardization, 2010).

As described in ISO 9241-210:2010, an HCD process consists of four activities:

- *Understand and specify the context of use:* Identify the users of the ship or system under development and determine their characteristics, goals, and tasks. Define the technical, organizational and physical environments in which the ship or system in question will operate.
- *Specify the user requirements:* Identify user needs and specify resulting functional and other requirements for the ship or system within the intended context of use.
- *Produce design solutions to meet user requirements:* Apply user requirements data and human factors considerations to produce design solutions and implement in the design.
- *Evaluate the design against the requirements:* Appraise the design solutions in order to select the most appropriate, identify defects and alter accordingly to produce the most optimal design.

Since the standards are not prescriptive rules but rather principles to follow, the actual activities and tasks, and the order of performance of these activities, as chosen by designers adopting HCD approaches, may vary greatly due to users' characteristics, intended operating environments, economic aspects and many other factors. HCD activities should be done whenever applicable and changed whenever necessary, e.g. one may evaluate a previous ship for better requirements or revisit the context of use for more detail on a design solution that is not working as well as it used to.

Standards, regulations and guidelines

The design and construction of ships and ship systems follow specific requirements, criteria, standards, rules, regulations, codes and guidelines. In the maritime industry, there are three primary sources from which those criteria can be obtained and applied: classification society rules, regulatory requirements and ship-owner requirements.

The first criterion to consider when starting a design project is the regulations and standards applied to the project outcomes. Standards and regulations set out the baseline for the minimum properties a designed object must possess to be able to enter into use in the industry. Failing to comply with such standards will render the project outcomes inadequate and jeopardize the whole design process.

Human factors consideration in the design of ships is officially required under the provisions of standards, rules and conventions issued by the International Maritime Organization (IMO), the International Labour Organization (ILO), the International Organization for Standardization (ISO) and the International Electrotechnical Commission (IEC). The IMO regulates human factors consideration in ship design through several assembly resolutions, resolutions and circulars from the maritime safety committee and the sub-committee on navigation, communications, search and rescue (NCSR), and the SOLAS Convention. The ILO Maritime Labour

Convention contains regulations addressing habitability and occupational health and safety. The IEC and ISO are two international organizations that produce international standards. IEC covers electrotechnology and related conformity assessment and ISO covers almost all other fields and conformity assessment with regard to vessel designs. Human factors dimensions are addressed in ISO/IEC standards of two fields of activity under the International Classification for

Table 7.4 IMO publications relevant to HF in ship design

Assembly Resolutions (RES)	
A.468(XII)	Code on Noise Levels on Board Ships
A.601(15)	Provision and Display of Manoeuvring Information on Board Ships
A.708(17)	Navigation Bridge Visibility and Functions
A.861(20)	Performance Standards for Shipborne Voyage Data Recorders (VDRs)
A.947(23)	Human Element Vision, Principles and Goals for the Organization
A.1021 (26)	Code on Alerts and Indicators
Maritime Safety Committee (MSC) Resolutions	
64(67)	Adoption of new and amended performance standards
128(75)	Performance Standards for a Bridge Navigational Watch Alarm System (BNWAS)
137(76)	Standards for Ship Manoeuvrability
190(79)	Performance Standards for the Presentation of Navigation-Related Information on Shipborne Navigational Displays
214(81)	Adoption of Amendments to the Performance Standards for Shipborne Voyage Data Recorders (VDRs) (Resolution A.861(20)) and Performance Standards for Shipborne Simplified Voyage Data Recorders (S-VDRs) (Resolution MSC.163(78))
232(82)	Adoption of the Revised Performance Standards for Electronic Chart Display and Information Systems (ECDIS)
MSC Circulars	
834	Guidelines for Engine-Room Layout, Design and Arrangement
982	Guidelines on Ergonomic Criteria for Bridge Equipment and Layout
1002	Guidelines on Alternative Design and Arrangements for Fire Safety
1091	Issues to be Considered When Introducing New Technology on Board Ship
1512	Guideline on Software Quality Assurance and Human-Centred Design for e-Navigation
Sub-Committee on Safety of Navigation (NAV) Circulars	
SN.1/Circ.265	Guidelines on the Application of SOLAS Regulation V/15 to INS, IBS and Bridge Design

Table 7.5 ISO documents relevant to HF in ship design

ISO	Title of document
2631-1:1997	Mechanical vibration and shock – Evaluation of human exposure to wholebody vibration – Part 1: General requirements
2923:1996	Acoustics – Measurement of noise on board vessels
3797:1976	Shipbuilding – Vertical steel ladders
5488:2015	Ships and marine technology – Accommodation ladders
5489:2008	Ships and marine technology – Embarkation ladders
20283-5:2016	Mechanical vibration – Measurement of vibration on ships – Part 5: Guidelines for measurement, evaluation and reporting of vibration with regard to habitability on passenger and merchant ships
17631:2002	Ships and marine technology – Shipboard plans for fire protection, life-saving appliances and means of escape
17894:2005	Ships and marine technology – Computer applications – General principles for the development and use of programmable electronic systems in marine applications
20283-2:2008	Mechanical vibration – Measurement of vibration on ships – Part 2: Measurement of structural vibration
20283-4:2012	Mechanical vibration – Measurement of vibration on ships – Part 4: Measurement and evaluation of vibration of the ship propulsion machinery
24409-1:2010	Ships and marine technology – Design, location and use of shipboard safety signs, safety-related signs, safety notices and safety markings – Part 1: Design principles
27991:2008	Ships and marine technology – Marine evacuation systems – Means of communication
8468:2007	Ships and marine technology – Ship's bridge layout and associated equipment – Requirements and guidelines
2412:1982	Shipbuilding – Colours of indicator lights
9241-210:2010	Ergonomics of human–system interaction – Part 210: Human-centred design for interactive systems

Standards (ICS), namely ICS.13 – Environment, Health Protection, Safety and ICS.47 – Shipbuilding and Marine Structures. Tables 7.4 and 7.5 provide lists of major publications from the IMO and ISO.

Alongside the above-mentioned organizations, classification societies provide guidelines to assist ship designers with the ergonomic design of ships and marine systems. American Bureau of Shipping (ABS), Lloyd's Register (LR), Det Norske Veritas Germanischer Lloyd (DNV GL), Bureau Veritas (BV) and the ClassNK (NK) are classification societies which represent the majority of the ship classification market. Each society develops their own guidelines addressing human factors in ship design. Their main guidelines are listed in Figure 7.6 together with the illustration of the involvement of different regulatory bodies in this issue.

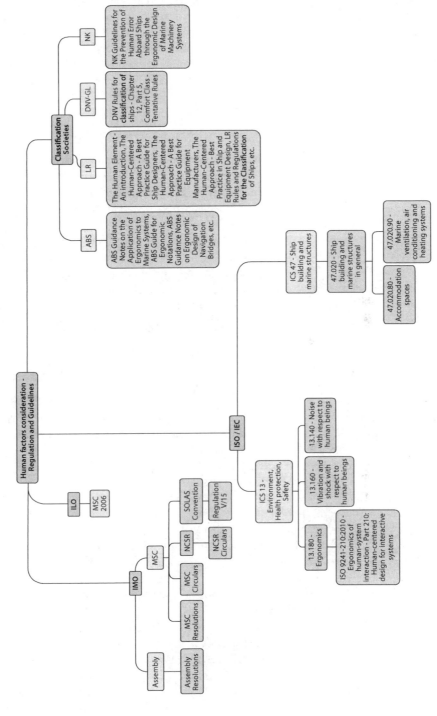

Figure 7.6 Human factors consideration in ship design – regulations and guidelines

The role of classification societies is expecting significant changes with the revolution in rule-making in the maritime industry by transforming from prescriptive regulations towards goal-based standards (GBS). With the introduction of GBS, the focus is shifted from the process to the achieved outcomes, where the designers have settled goals but are free to elect whatever measures they deem fit to achieve compliance. The structure for goal-based new ship construction standards have explicit consideration of the human element, as seen in the following extract from the resolution MSC.287(87) (IMO, 2010, p. 6):

> Ship's structures and fittings shall be designed and arranged using ergonomic principles to ensure safety during operation, inspection and maintenance. These considerations shall include, but not be limited to, stairs, vertical ladders, ramps, walkways and standing platforms used for means of access, the work environment, inspection and maintenance and the facilitation of operation.

The move towards goal-based standards has the advantage of being highly flexible to accommodate not only human variability and adaptability but also the dynamics of the marine system due to changing technology, operations and manning. However, this novel approach also requires changes in the practice of many stakeholders in the industry. Explicit activities are expected from classification societies throughout the design, construction, survey and approval to demonstrate compliance (Earthy & Sherwood Jones, 2006). Within the framework set at the IMO, it is the role of classification societies to develop specific criteria to support the goals.

Challenges with human-centred design

It is undeniable that human-centred design helps create good products, but just like with any other design approach, there are advantages and disadvantages. The important thing is that the designers understand this and adapt in order to find the most appropriate design practice.

Applying knowledge of human factors and ergonomics will improve usability, reduce the likelihood of human errors and enhance system effectiveness and efficiency. However, there are aspects of human-centred design which designers must take into consideration. One drawback of HCD is the reallocation of time and expense. The process of user data collection, subsequent analysis and producing design solutions requires both specialist human and financial resources. It incurs some short-term risk to project time and cost. However, this is more than repaid in reduced long-term risks to safety, efficiency and training costs. This issue will be further discussed in the section on the economic aspects of design.

An important aspect of HCD is the involvement of users in the design process. Users may not be aware of their needs or make false assumptions of their actual needs. Furthermore, acceding to users' requests instead of finding out what they need may result in overly complex designs. Users are not designers and their involvement should reflect this distinction. A common mistake made by designers, when they do involve end users, is to utilize users as co-designers. It is important to bear in mind that 'HCD does not necessarily imply that users perform research or design roles ... but that they contribute to research or design processes, as experts on their own daily lives and on their own experiences with products and services' (Steen, 2008, p. 27).

The economic aspects of design

Another issue when it comes to the implementation of human factors considerations is the monetary challenge. Since such design practice may require additional expense in the early phases, it should be considered an investment, and just like any investment, the managers always pay keen attention to economic justification. However, it seems there is still a lack of awareness in the industry regarding the economic benefits that optimal design may bring in the long run. This may be due to a lack of routines and methods for performing cost and effect estimations of ergonomics investments (Österman, 2009, 2012; Österman & Rose, 2015).

Costa and Lützhöft (2014) analysed and categorized the benefits of addressing human factors in ship design and operation. Their study found that implementing HF in the design stages can result in not only direct benefits for the seafarers but also several benefits for the ship owners. Among them, the following economic benefits can serve as a strong justification for the ship owners to invest in HF:

- Reduction of accidents leads to reduction of operational costs
- Less additional cost for adjustment or redesign since everything was done correctly at the beginning
- Prevention of layday cost due to crew mistakes
- Less stressful working environment results in less sick leave
- Lower training costs if systems are more intuitive
- Lower costs for relocation of hardware if the system is built properly from the start
- Better public image and reputation for the company, which can also lead to economic benefits.

There is also an emphasis on the importance of early application of good design practices. Yet, making such investing decisions can be problematic due to the fragmented structure of the maritime industry, where one department pays for the design and another reaps the benefits of less maintenance, less attrition, safer and more efficient operations. Unfortunately, there are few examples of cost-effectiveness calculations available for the shipping domain. However, we can assume that if we wait until we are at operational and management levels to address human factors issues we tend to increase redesign costs, training costs and paperwork (e.g. ISM). An example from BMT regarding an oil production company's benefits from integrating human factors and engineering early into design of offshore facilities identifies the following reductions (BMT, 2007):

- Capital expenditure 0.25–5 per cent
- Engineering hours 1–10 per cent

- Design rework 1–5 per cent
- Project duration up to 40 per cent less due to reduced rework and fewer approval cycles
- Operation and maintenance 3–6 per cent per year.

Despite the current lack of a formal method for evaluating economic benefits of design investment in the industry, it is suggested that optimal design can improve user performance and well-being (Petersen, 2010) and ultimately lead to improved safety.

If people consider the potential cost-reduction and cost-effectiveness behind the safe operation of ships and shipboard equipment, safety can be perceived as profit, as added value and as a corporate social responsibility which must be considered when making investment decisions.

Human factors and design engineering

Through their active engagement in the design process, naval architects have a direct influence on every aspect of ships' designs. While their work is influenced by rules and guidelines from regulatory bodies, it is the professional knowledge and maritime HF skills of the designers that are essential for the design of ships and marine systems with high usability. Unfortunately, designers are usually biased towards the technical aspects and focus exclusively on the ship as a physical artefact while ignoring work design and human–system integration. It is necessary that maritime designers become more aware of this issue and employ design approaches that can employ HF concepts to their full extent (Lützhöft, Petersen, & Abeysiriwardhane, 2017). To achieve this, we first need to uncover the reasons behind this phenomenon.

The engineering mindset

Efforts have been made to intervene early in the process, at ship design level (Abeysiriwardhane et al., 2014; Abeysiriwardhane, Lützhöft, Petersen, & Enshaei, 2015), and at software/equipment level (Petersen, 2010, 2012). The former is a summary of a project on increasing HF awareness among the community of naval architects, and the latter is an account of an in-house project using HCD to design and test a new generation of maritime electronics. Both used the ISO standard as their starting point. One of the issues found in both studies is the differing underlying philosophy, culture and mindset in human factors and engineering.

The lack of value assigned to human factors in the professional viewpoint held by naval architects and maritime designers has its roots in the natural inclinations of design engineering, the training and education systems, and design engineering thinking (Lützhöft et al., 2017).

In his theory of multiple intelligences, instead of seeing intelligence as dominated by a single general ability, Gardner (1998) differentiates intelligence into eight different abilities. The archetypical design engineer is gifted with four of these, namely: naturalistic intelligence, spatial intelligence, bodily kinaesthetic intelligence and logical-mathematical intelligence. Naturalistic intelligence represents the abilities to recognize and categorize natural objects; spatial intelligence allows people to form and operate on mental images; bodily kinaesthetic intelligence provides the capability to control body motion to handle objects; and logical-mathematical intelligence allows the ability to confront and assess objects and abstractions and understand their relations and underlying principles (Gardner, 2006a, 2006b). In combination, these traits provide a person with the necessary skills to design and construct objects, but also lead to the design thinking that is based on numbers and logic. Thus, engineering is based on applied mathematics and physics and their associated principles, which explains why typical engineering education programmes often focus on technological subjects rather than the issues of human–machine interaction (Lützhöft et al., 2017). In the first study mentioned above, Abeysiriwardhane et al. (2014) study undergraduate naval architecture students at the Australian Maritime College and find the students lack awareness in maritime human factors issues. None of the participating students was aware of HF rules and guidelines even though most of them were familiar with marine regulations and classification societies.

Furthermore, the nature of design engineering thinking urges designers to approach complex affairs by dividing them into semi-independent components (Bucciarelli, 1994). The success of this strategy depends upon suppressing potential issues through the avoidance of qualitative requirements and relying solely on measurable physical criteria. Also, uncertainty is unavoidable during the process of creating a new product, and designers have always faced the risk of uncertain knowledge. Through the years, design engineers have developed strategies to tackle uncertainty; they often rely on immediate data and employ heuristics rather than comprehensive analysis to even out the lack of precise information (Lützhöft et al., 2017). Together, these strategies make it problematic for designers to consider and apply knowledge on the context of use and user characteristics in their designs.

To address the differing cultures of Human Factors and technical factors, it is necessary to recalibrate the engineering mindset, which ought to balance the requirement of technical functions and the needs and satisfaction of the end users. This re-calibration has to be done in the training of future designers, by integrating knowledge of human factors in the curriculum. However, it is clear that one-sided change does not work. Thus, the human factors professional must adapt the content and dissemination techniques to practice, and not preach. For undergraduate education, it is recommended that the following key areas are addressed in the syllabus (Abeysiriwardhane et al., 2014):

- Fundamental principles of HF and HCD
- Development of an understanding of the nature and application of these principles
- Guidance on HF rules and regulations followed by the industry
- Designers' responsibility to integrate HF and HCD into designs to avoid hazards

- Appreciation of the HF and HCD in complex work systems
- Demonstration of the importance for safety and comfort for the life of the end users
- How to become a usability expert in a working environment where other employees are not aware or interested in HF/HCD.

Interestingly, when attempting to involve engineers in designing ships and ship systems for seafarers, both authors of the two studies mentioned in the beginning of this section come to the same conclusion – that practice bests theory. Instead of the traditional teaching methods, it is better to expose students to the real environment of the end users, and to provide them with first-hand experience on the life and work on board ships. This new approach has been shown to develop the awareness of maritime HF among students and improve the connection between the future designers and the end users (Abeysiriwardhane, Lützhöft, Petersen, & Enshaei, 2016).

Basically, if human factors knowledge and skills are to be accepted in technical domains, they may have to be reframed to fit the 'end user' – couched in a language that makes sense to technical users and presented in a manner which fits the engineering mindset.

Ergonomics mindset at the organizational level

Achieving human factors awareness at an individual level is only the first step of the 'Design for safety' approach. To achieve safety through design at the industry level, we need to rebalance the engineering mindset in every organization, and especially for the design of ships and shipboard equipment, i.e. in ship design firms, shipyards, and equipment manufacturers. The goal of this transition is to make system usability and human factors a focus in each organization's operational policies.

One useful instrument for this process is the Organizational Human-Centredness Scale, introduced by Lloyd's Register, which describes levels of usability maturity (Table 7.6) – the extent to which an organization addresses human-centred issues.

The descriptions of attributes at each maturity level explain why human-centred methods developed for high-level organizations are not applicable for lower level organizations; also, low-level organizations cannot immediately produce solutions to achieve the highest level of maturity but rather must undertake a steady process of transition through all levels.

The main usage of the scale is to evaluate and improve an organization's capability in addressing human factors and, in due course, improve the quality of use of that organization's products. However, besides the primary use in assessment, the scale can also be used as a model for designing an organization's operations or to improve existing operational agendas. For the marine sector the scale has been developed into the referenced LR Guides (Lloyd's Register, 2007, 2014a, 2014b) which provide marine-specific capability levels for HCD process improvement for operators, designers and manufacturers.

Table 7.6 Usability maturity levels and process attributes

Level	Description of Levels	Process Attributes
Level X	Unrecognized The need for a human-centred process is not recognized.	No indicators There is no concern with mixed level of user satisfaction. There are no positive attributes at this level.
Level A	Recognized The organisation recognizes the need to improve the quality of use of its systems.	A1 – Problem recognition attribute Members of the organization are aware of the problem with the quality of use of the systems produced. A2 – Performed processes attribute Process are performed to provide input to make the system human-centred.
Level B	Considered The organization engages in training and awareness raising to make its staff aware that quality of use is important and can be improved by taking account of end-user requirements during the development process.	B1 – Quality in use awareness attribute The staff involved in the development process are aware of quality of use as an attribute of the system. B2 – User focus attribute The staff involved in the process relating to the user-facing element of the system take account of the fact that a human being will need to use it.
Level C	Implemented Human-centred processes are fully implemented and produce good results.	C1 – User involvement attribute Information is collected from representative users using appropriate techniques throughout the system life cycle. C2 – HF technology attribute Human factors methods and techniques are used in or by human-centred processes. C3 – HF skills attribute Human factors skills are used in human-centred processes.
Level D	Integrated Human-centred processes are used in the improvement of work products from other processes.	D1 – Integration attribute Human-centred processes are integrated into the quality process and systems life cycle of the organization. D2 – Improvement attribute Human-centred process are integrated with other processes. D3 – Iteration attribute The development lifecycle is iterative.
Level E	Institutionalized The quality in use of whole ranges of systems is coordinated and managed for business benefit. The culture of the organisation become benefits from being user and human centred. Human-centred skills are regarded on a par with engineering skills.	E1 – Human-centred leadership attribute Human factors/people-centred approach influences the management of all systems life cycle processes. E2 – Organizational human-centeredness attribute Human factors/people-centred approach influences the attitude of the whole organization.

This scale is just one of several methods to expand human factors considerations at an organizational level; other methods such as the Corporate UX Maturity Scale (Nielsen, 2006), where the level of commitments to user experience (UX) of an organization is sorted in eight stages ranging from objecting to fully committed, can also be considered. In practice, each organization should follow the most appropriate approach, whilst bearing in mind the goal of usability and user needs.

Conclusions and recommendations

The properties of a product are defined by its design. The design of ships and equipment will directly affect the outcomes of shipboard operations and thus determine maritime safety. Within the scope of maritime shipping, the characteristics of ships and shipboard equipment are determined by naval architects and equipment manufacturers during the design process.

Investigation reports of maritime accidents indicate that the main reasons behind the majority of accidents are inappropriate design factors resulting in poor usability, impeding proper interactions between the human and other elements of the maritime system as well as increasing the probability of human error. Such faulty designs may be the result of a lack of proper understanding about the capabilities and behaviours of the actual users among the naval architects and equipment designers' communities.

To improve safety, it is necessary to design ships and ship equipment with due regard to human factors in order to match with users' abilities and requirements. This practice is, however, often neglected in the maritime industry. There is a need for a systematic approach for dissemination of this issue to all stakeholders in the industry as well as a request, and ideally a requirement, for all relevant parties to take action to follow human factors design practice. In establishing such an approach, the following factors have to be considered:

- Design solutions that consider the human element are developed by following human-centred design approach to provide design solutions based on users' needs in the intended context of use.
- Project managers, naval architects and equipment designers should take account of the changes to resource and management requirements associated with the introduction of human-centred design activities to a project.
- Equipment designers and naval architects need to follow required practices prescribed in regulations and standards as well as refer to guidelines from classification societies for recommended applications.
- Besides safety, human-centred design also provides an economic benefit. An optimal design will increase effectiveness, efficiency, productivity and reduce maintenance cost. Furthermore, it is important to acknowledge that higher safety can also be perceived as monetary profit.
- Design principles need to be rebalanced between human and engineering factors, with stronger consideration placed on the human element. This issue needs to be addressed at the individual level through the incorporation of

HF knowledge in the training syllabus for future industry designers, and at the organization level by enhancing the level of user-centredness among the industry's organizations.

With proper understanding of the importance of the design process to safety and productivity, particularly as seen in recent design projects, we can expect a transformation in the practice of ship and marine equipment design in the near future, which will strengthen the contribution of design to safe and effective operations in the maritime industry.

References

Abeysiriwardhane, A., Lützhöft, M., & Enshaei, H. (2014). Human factors for ship design; exploring the bottom rung. *International Journal of Marine Design*, 156(C1), 153–62.

Abeysiriwardhane, A., Lützhöft, M., Petersen, E., & Enshaei, H. (2016). Incorporate good practice into ship design process; future ship designers meet end users. Paper presented at the ERGOSHIP 2016: Shaping Shipping For People, Melbourne, Australia, 6–7 April.

Abeysiriwardhane, A., Lützhöft, M., Petersen, E. S., & Enshaei, H. (2015). Future ship designers and context of use: Setting the stage for human centred design. Paper presented at the Marine Design, 2–3 September, London.

Alert! Project (2006). *International Maritime Human Element Bulletin*, 11.

Alert! Project (2014). *International Maritime Human Element Bulletin*, 34.

Alert! Project (2015a). *International Maritime Human Element Bulletin*, 37.

Alert! Project (2015b). *International Maritime Human Element Bulletin*, 38.

Alert! Project (2016). *International Maritime Human Element Bulletin*, 40.

American Bureau of Shipping (2006). *Guide for Vessel Maneuverability*. Houston, TX: American Bureau of Shipping.

American Bureau of Shipping (2016). *Guide for Crew Habitability on Ships*. Houston, TX: American Bureau of Shipping.

Australian Transport Safety Bureau (2015). *Unintentional Release of the Freefall Lifeboat from Aquarosa, Indian Ocean on 1 March 2014*. Canberra: Australian Transport Safety Bureau.

Bialystocki, N. (2016). Human factors dimensions in the design of a PCTC vessel. Paper presented at the RINA Human Factors in Ship Design and Operation Conference, London, 28–29 September.

BMT (2007). *Human Factors Engineering Success Stories in the Oil and Gas Industry*. Alexandria, VA: BMT Designers and Planners

Bucciarelli, L. L. (1994). *Designing Engineers*. Cambridge, MA: MIT Press.

Chaplin, N., & Nurser, J. (2007). Launching the Tamar. *Ingenia*, 33, 37–43.

Costa, N., & Lützhöft, M. (2014). The values of ergonomics in ship design and operation. Paper presented at the RINA Human Factors in Ship Design and Operation Conference, London, 26–27 February.

Dhillon, B. S. (2007). Human error in shipping. *Human Reliability and Error in Transportation Systems* (pp. 91–103). London: Springer London.

Earthy, J. (1998). *Usability Maturity Model: Human Centredness Scale*. INUSE Project Deliverable D, 5, 1–34. Teddington: Information Engineering Usability Support Centres

Earthy, J., & Sherwood Jones, B. (2006). Design for the human factor: The move to goal-based rules. WMTC: 2nd World Maritime Technology Conference: Maritime Innovation, Delivering Global Solutions, London, 6–10 March.

Earthy, J., & Sherwood Jones, B. (2011). *Best Practice for Addressing Human Element Issues in the Shipping Industry.* London: Lloyd's Register.

Earthy, J., Sherwood Jones, B., & Squire, D. (2016). Improving awareness of the human element in the maritime industry. *Alert!* http://www.he-alert.org/objects_store/Alert_Issue_1.pdf

Ebeling, C. E. (2004). *An Introduction to Reliability and Maintainability Engineering.* New York: Tata McGraw-Hill Education.

Equasis (2015). The world merchant fleet in 2015 statistics from Equasis. Available online at: http://www.emsa.europa.eu/equasis-statistics/items.html?cid=95&id=472

European Maritime Safety Agency (2016). *Annual Overview of Maritime Casualties and Incidents 2016.* Lisbon: European Maritime Safety Agency. Available at: http://www.emsa.europa.eu/news-a-press-centre/external-news/item/2903-annual-overview-of-marine-casualties-and-incidents-2016.html

Gardner, H. (1998). A multiplicity of intelligences. *Scientific American,* 9(4), 19–23.

Gardner, H. (2006a). *Changing Minds: The Art and Science of Changing Our Own and Other People's Minds.* Cambridge, MA: Harvard Business Review Press.

Gardner, H. (2006b). *Multiple Intelligences: New Horizons.* New York: Basic Books.

Grech, M., Horberry, T., & Koester, T. (2008). *Human Factors in the Maritime Domain.* Boca Raton, FL: CRC Press.

International Ergonomics Association (2012). What is ergonomics? http://www.iea.cc/ergonomics.

International Labour Organization (2006). *Maritime Labour Convention.* London: IMO.

International Maritime Organization (2003). *A.947(23) Human Element: Vision, Principles and Goals for the Organization.* London: IMO.

International Maritime Organization (2010). *MSC.287(87): Adoption of the International Goal-Based Ship Construction Standards for Bulk Carriers and Oil Tankers.* London: IMO.

International Organization for Standardization (2010). *Ergonomics of Human System Interaction – Part 210: Human-Centred Design For Interactive Systems* (ISO 9241-210:2010). Geneva: International Organization for Standardization.

Lloyd's Register (2007). *The Human Element: Best Practice for Ship Operators.* London: Lloyd's Register.

Lloyd's Register (2008). *The Human Element: An Introduction.* London: Lloyd's Register.

Lloyd's Register (2014a). *The Human-Centred Approach: A Best Practice Guide for Equipment Manufacturers.* London: Lloyd's Register.

Lloyd's Register (2014b). *The Human-Centred Approach: A Best Practice Guide for Ship Designers.* London: Lloyd's Register.

Lützhöft, M. (2004). The technology is great when it works: Maritime technology and human integration on the ship's bridge. PhD thesis, University of Linkoping.

Lützhöft, M., Petersen, E. S., & Abeysiriwardhane, A. (2017). The psychology of ship architecture and design. In M. MacLachlan (ed.), *Maritime Psychology: Research in Organizational and Health Behavior at Sea* (pp. 69–98). Cham: Springer International Publishing.

Macrae, C. (2009). Human factors at sea: Common patterns of error in groundings and collisions. *Maritime Policy and Management,* 36(1), 21–38.

Marine Accident Investigation Branch (2001). *Review of Lifeboat and Launching Systems' Accidents*. Southampton: Marine Accident Investigation Branch. https://www.gov.uk/government/publications/lifeboats-and-launching-systems-accidents-review

Marine Accident Investigation Branch (2008). *Report on the Investigation of the Grounding of CFL Performer Haisborough Sand North Sea 12 May 2008*. Southampton: Maritime Accident Investigation Branch.

Marine Accident Investigation Branch (2014). *Report on the Investigation of the Grounding of Ovit in the Dover Strait on 18 September 2013*. Southampton: Marine Accident Investigation Branch.

Miller, G. (2000). The current and suggested role of human and organizational factors (HOF) for the reduction of human induced accidents and incidences [sic] in the maritime industry. Paper presented at Maritime Human Factors 2000, Linthicum, MD.

Myles, H. (2015). The human element in ship design. *Alert!* http://www.he-alert.org/filemanager/root/site_assets/standalone_article_pdfs_1220-/he01365.pdf

National Transportation Safety Board (2014). *Collision of the Passenger Vessel Seastreak Wall Street with Pier 11, Lower Manhattan, New York, New York January 9, 2013*. Washington, DC: National Transportation Safety Board.

Nielsen, J. (2006). Corporate usability maturity. http://www.useit.com/alertbox/maturity.html

Österman, C. (2009). Cost-benefit of ergonomics in shipping. Master's thesis, Linköping University.

Österman, C. (2012). *Developing a Value Proposition of Maritime Ergonomics*. Gothenburg: Chalmers University of Technology.

Österman, C., & Rose, L. (2015). Assessing financial impact of maritime ergonomics on company level: A case study. *Maritime Policy and Management*, 42(6), 555–70.

Petersen, E. S. (2010). User centered design methods must also be user centered: A single voice from the field. A study of user centered design in practice. Licenciate thesis, Chalmers Technical University, Gothenburg.

Petersen, E. S. (2012). Engineering usability. PhD thesis, Chalmers University of Technology, Gothenburg.

Ralph, P., & Wand, Y. (2009). A proposal for a formal definition of the design concept. In K. Lyytinen, P. Loucopoulos, J. Mylopoulos, & B. Robinson (eds), *Design Requirements Engineering: A Ten-Year Perspective: Design Requirements Workshop, Cleveland, OH, USA, June 3–6, 2007, Revised and Invited Papers* (pp. 103–36). Berlin: Springer.

Reason, J., & Hobbs, A. (2003). *Managing Maintenance Error: A Practical Guide*. Boca Raton, FL: CRC Press.

Roberts, S., Nielsen, D., Kotłowski, A., & Jaremin, B. (2014). Fatal accidents and injuries among merchant seafarers worldwide. *Occupational Medicine*, 64(4), 259–66.

Rothblum, A. (2000). Human error and marine safety. Paper presented at the National Safety Council Congress and Expo, Orlando, FL.

Royal National Lifeboat Institution (2014). SIMS: Technology that puts crew safety first. *RNLI Magazine*. Retrieved from http://magazine.rnli.org/Article/SIMS-Technology-that-puts-crew-safety-first-4

Sherwood Jones, B. (2016). Regulation of occupational health and safety: A human factors viewpoint. Paper presented at the Seahorse International Conference on Maritime Safety and Human Factors, Glasgow 21–23 September.

Smith, D. J. (2011). *Reliability, Maintainability and Risk: Practical Methods for Engineers Including Reliability Centred Maintenance and Safety-Related Systems*. Amsterdam: Elsevier.

Steen, M. (2008). *The Fragility of Human-Centred Design*. Amsterdam: IOS Press.

Transport Accident Investigation Commission (2011). *Final Report: Marine Inquiry 11-201, Passenger Vessel Volendam, Lifeboat Fatality, Port of Lyttelton, New Zealand, 8 January 2011*. Wellington, New Zealand: Transport Accident Investigation Commission.

8 Autonomous ships, ICT and safety management

Jonathan V. Earthy and Margareta Lützhöft

Background

Technology is being promoted by several original equipment manufacturers (Rolls-Royce, 2016; Wärtsilä Corporation, 2017). BHP Billiton has a vision of the future of dry bulk shipping that comprises safe and efficient autonomous vessels carrying BHP cargo, powered by BHP gas (Bhatti, 2017). The company is seeking partners with which to work on the technology and believes it may happen within a decade.

Increased supervision from the shore is another theme. TRANSAS says that ECDIS (Electronic Chart Display and Information System), ship stability, voyage planning and weather forecast data as well as fuel consumption, engine performance and bridge alarm system data should be made available in real time to the shore as well as on the ship (TRANSAS, 2016). Maersk has indicated that it may be looking at some form of autonomous ships in the 2030–35 timeframe, which would coincide with the end of the useful life of their recent new builds (Quick, 2016).

Not all organizations are convinced, warning that the 'Silicon Valley' obsession with technology may not be a cost-effective approach. The classification society ABS says it is seeing no demand for such ships from owners, and is instead focusing on technological developments for the world's existing fleet. Jan O. De Kat, Technical Director of global marine for ABS, says 'We have not seen any company that says we want to make our ships autonomous'. This is backed up by Egil C. Legland, country manager of ABS Norway, who says 'We need to ignore the hype that is going on at the moment and focus on the reality.' They believe that autonomous ships will take a long time to develop fully and that there are issues important to ship owners that need to be addressed in the mean time. Examples quoted are costs, productivity and creating value. Legland further says that ship owners will not push for autonomy without a clear strategic or financial advantage (Hand, 2017).

Whether the industry adopts autonomous (i.e. self-determined) systems controlling whole ships in the near future is uncertain. However, parts of platforms are already under offship supervision or completely automated. Furthermore, there is increasing use of Information and Communication Technology (ICT) for a range of reasons such as:

- regulatory demand
- reduced experience onboard
- technical complexity associated with increased performance
- technical complexity acquired accidentally with equipment/systems
- enterprise resource planning (ERP)/administration software
- systems to provide shore with more monitoring, advice and control
- communications.

Any or all of these may have safety effects from concealing, misinforming or overloading with information and this could have an adverse influence on command decisions at all levels in the operational hierarchy of the ship as a tool for business or as a technical platform.

There is frequently not a conscious intent to acquire complex ICT, so the unexpected costs and responsibilities associated with ownership may not be managed. For example, no formal training in using a system is provided after the installers brief the first crew; or, owners are unwilling to purchase the data analysis modules that help crew understand/make the system useful, safe and fully meet the intent of computerising the system in the first place.

Perhaps the most widely used argument for autonomous/unmanned ships is safety. For, example the claims from DNVGL (DNV GL, 2014, p. 39) and also NUMAST (Graveson, 2015) that reducing manning through using technology removes human error and difficult, dangerous and dirty jobs from the industry. There are also more extreme claims asserting a moral imperative to use artificial intelligence (AI) to reduce accident rates.

The assumption that people do more harm than good also needs to be challenged. There is increasing evidence (or perhaps it is just a new perspective) that humans add safety to systems by adapting and thus mitigating risks; effectively, humans are 'making safety'. It is an idea following from the development of Safety-II (discussed in Chapter 1), the resilience engineering community (Hollnagel, 2014) and from a pioneer in safety theories, Reason (2008). In analysing these claims in a maritime context, we first should differentiate between risk to people's safety and risk to safety of the assets (ship, cargo and reputation). People are there as workers to protect the safety of the asset, originally to keep it going (less so now), to provide flexible support in emergencies (fire, etc.), to mend things (engineering, information technology), and at an executive level to detect defects and make strategic decisions (both long-term and immediate) on sometimes limited data, but supported by experience.

A superficial reading of IMO (International Maritime Organization) A.1047 Minimum Safe Manning suggests that a reduction in manning through automation is an allowed derogation with no stated caveats about usability. However, the more detailed analysis presented later in this chapter does not accord with this view. As autonomy advances into intelligent, self-learning systems, underpinned by artificial intelligence, assuring safety will become more challenging. For marine regulation, demonstration through evidence and experience will be important.

Approach followed in this chapter

This topic area is evolving very rapidly. This chapter is a report from the front line and as such relies heavily on a few recent sources, some of which have not yet been reported in a formal manner. The aim is to present a kit of parts for consideration of the issues associated with the safety of operations based on advanced ICT rather than a clear set of recommendations for ensuring safety in this new world. We wish to make it very clear that this chapter presents the personal observations of the authors and is not the approved position of their employers or other affiliations. References in this chapter are used not as justifications but as links to further information.

The authors wish to attribute and acknowledge the primacy and value of their sources and how they have been used. Several of these will be difficult to access and this is why we quote extensively from work by Lloyd's Register and the Southampton Marine and Maritime Institute. This includes the findings of joint workshops on marine ICT organized by Shenoi and Earthy in 2016. We use abstracted recommendations from Veal and Tsimplis (2017). The Minimum Safe Manning section contains personal interpretations. Sherwood Jones and Earthy (2016) provide much of the intellectual underpinning of the chapter. The section on human factors assurance for advanced automation is derived from Lloyd's Register's ongoing work in this area (Berry, Stokes, & Toomey, 2016; Lloyd's Register, 2016b).

Contents and use of this chapter

How will you be meeting your regulatory responsibility for the ship? Remember, there are portions of the platform that already behave in an autonomous fashion. We already have some decision-making that is completely out of the crew's hands: for example, engine governors and automatic update of navigational charts. However, you do not need to wait for autonomy to have problems with human factors and poor usability as these can be found anywhere where ICT is used. This causes concern for safety, and human-centred design is not optional.

1   We want to give clues to what should be added or considered for ISM (International Safety Management). A starting point for extending ISM audits to incorporate human factors is to use the Maritime and Coastguard Agency (MCA) human element assessment tools (HEAT) for operational issues (Maritime and Coastguard Agency, 2008).
2   By changing the technology, you may need to change your Minimum Safe Manning (MSM), especially if technology allows you to do things differently from shore – your minimum safe manning declaration should say why it is safe to do this. So it needs to include the shore staff – because your system extends beyond the ship. For remote control the question you need to answer is *why is safe manning on board zero?*

On a cyber-enabled ship, ICT informs, assists or even replaces the traditional seafarer and the shore-based tasks that keep the ship operational and safe, and carry out the business of the owner. This has potentially great benefits, but these can only be realized if the ship's design addresses the human–system issues that emerge from using ICT. It is not enough to consider the user interface for each piece of technology in isolation (Lloyd's Register, 2016a).

For dependability and reliability, the design, construction and management of ICT has to enable both onboard and remote operators to work safely, securely and effectively. We wish to see which of these issues a safety management system can address. First, according to the International Safety Management code, and secondly in the implementation of it into individual Safety Management Systems – will it be done if it is not mandatory?

Specifically, Table 8.1 outlines the main human element challenges related to widespread and intensive use of marine ICT.

Addressing these challenges requires a structured, human-centred approach to system development and operation in combination with a consideration of ICT and its management within the SMS and other elements of the regulatory framework. Implementation of human-centred design is described in Chapter 7 of this book and Lloyd's Register (2014a, 2014b).

What are the issues associated with advanced ICT?

This section sets a context by presenting the range of concerns and hazards associated with the increasing use of information and communication technologies

Table 8.1 Human element issues for ships using advanced ICT

	Issue	Addressed by ISM/SMS
1	Design of equipment and systems needs to consider the changed expectations placed on the users to operate them and diagnose failures and other problems.	Not for the ship proper but the ship system, see Circ. 1512
2	The jobs of seafarers and shore staff need to be redesigned to take account of new or changed responsibilities, including support and maintenance of software-intensive systems.	To some degree, if procedures can be said to be job descriptions, STCW[1] is a basis
3	Numbers, roles, skills and competence needs to be assessed to determine safe manning requirement.	Already covered by A.1047 minimum safe manning
4	The cumulative effect of all changes on the safe and effective performance of seafarers and shore staff needs to be considered, in terms of situational awareness and training.	Not much help. ISM and STWC should address this
5	Ship operations need to be monitored to ensure that the human component of the systems is performing safely and effectively and that adequate maintenance is being carried out.	Procedures can assist well here. ISM should address this

Note
1 The International Convention on Standards of Training, Certification and Watchkeeping for Seafarers)

(ICT) in the marine industry. The human factors issues associated with automation are then discussed.

In 2016 as part of an initiative to establish a common understanding amongst a wide range of stakeholders in shipping regarding the opportunities and challenges associated with the use of ICT, the Southampton Marine and Maritime Institute and Lloyd's Register (under the leadership of Professor Ajit Shenoi and Earthy) held workshops to review the benefits and issues associated with the marine use of ICT. The concerns identified at these events are outlined below.

Skill set

Is there a need to change this on-board ships? Engineers may not have the required skills to maintain future platforms, more instruction from shore will be commonplace, more traditional knowledge may be lost/replaced by ICT and the quality of training may decrease.

Reliability and verification

ICT systems in combination may act unpredictably, more so than the systems they may be controlling, e.g. engines. The aviation industry, for example, has several failsafe systems – is this level of integrity necessary for shipping? (A dead ship is less of a problem than a dead plane.) Complete system failures are possible as well as a lack of willingness to discuss and address reliability. Marine software engineering is not at the same level of quality as other fields of engineering.

In discussing all marine ICT it was identified that there is a broad range of quality and there are significant concerns about the lack of established software engineering best practice in the development of systems for management and reporting applications. The fast pace of change in ICT systems combined with a lack of validation and verification may lead to quality issues – an 'attitude' of patching rather than initial quality is commonplace. This introduces risk, particularly for business-critical systems.

Overdependence

Systems offer improved situational awareness (e.g. AIS (Automatic Identification System)) but the risk of over-reliance and reduced vigilance is present. Using ICT to augment rather than replace competence can assist assimilation of information, particularly in task-oriented environments (e.g. data fusion). Future seafarers are a generation of 'digital dependents' and this brings issues of 'blind trust' when using systems, even though understanding whether an output is appropriate/abnormal is crucial.

ICT systems can introduce complacency – engine control rooms are filled with 'alarms' and 'beeps' and engineers accept this rather than demand good alert design and management as the norm.

Reduction/introduction of risk

Transferring to new vessels, preparing/handling rare situations or incidents: is the risk associated with ICT systems accepted/understood by those using it? There is currently no definition of an ICT near miss and so understanding and reporting are difficult. There is also a reputational risk associated with non-reporting, e.g. not complying with European requirements to report cyber attacks. Ransomware is possible, and there is a lack of competence to deal with this and other such incidents.

Security

This is a major area of focus in ICT. Viruses and malware can increase costs and compromise operations, particularly when crew members have open access to internet/systems (e.g. using USB ports on bridge equipment to charge a phone). Constant connection to the internet means security updates become a requirement. Rolling this out means additional costs and a need for constant support from the operator's IT department and the supplier.

There is a belief that there is a difference between shore-side and ship security, with the latter being modest in comparison with the former – the ship is still seen as 'isolated', when in reality it is very much connected. There must be a culture change to promote understanding of this, through quality education and training.

Pace of change

ICT development and change is fast, in contrast to the shipping industry which is largely conservative, reactive and slow. The fast pace of change may also lead to quality and reliability issues. New features/updates are being added and removed constantly so designers remain competitive but ship owners want a single product that works. Being forced to update software is not in their interests, particularly if the current version works.

Maintainability

Maintaining ICT-based systems requires a different skill set to that of a Chief Engineer, therefore ship operators may become reliant on suppliers rather than themselves/crew to maintain the systems and diagnose issues. If they are unwilling to purchase diagnostic software, in the event of a failure the supplier must be contacted, leading to increased costs and loss of control.

Differences in coding practices mean that interrogation of systems is often difficult and slow, and therefore again requires contact with the supplier to resolve issues and maintain systems.

Roles of ship and shore

Seafarers may feel that shore teams do not fully understand how a ship operates, e.g. separate shore units asking for the same information. Crew may feel devalued if too much responsibility is given remotely to shore. Additionally, when decisions are to be made using ICT systems, should they offer decision support or decision management/taking? Rules and procedures are necessary and research is needed into how to manage a shift of control, or how a 'hybrid control' can be established.

There is also the issue of more commercially interested intrusion from shore-side, and lack of trust towards seafarers. Constant monitoring by shore operations increases transparency but may give crew the feeling that they are disconnected/lack control/being watched by 'big brother', e.g. fuel usage monitoring.

Usability of systems

Many of the ICT systems used on board ships are not designed with ships in mind. They are developed by non-maritime suppliers, without the involvement of mariners, and hence the software is not designed for the job. There is a lack of common User Interface, and often a lack of understanding about how to interpret the data received. Too much variation in software design causes issues, such as when crew transfer to a new ship (training is required to learn new systems). The assumption that people are technologically able is one that is incorrect, but changing with the new generation of seafarers. Being able to use a computer is different to using it to do a job effectively.

Collaboration and the achievement of best practice is needed. The challenge lies in developing solutions which work for everyone but recognize differences in users and situational requirements.

Ownership of data

Large amounts of data are produced by ships but who owns these? Who is responsible for the security of these? A solution is to understand where sharing and access differ, and to whom these apply.

Autonomy

This introduces massive challenges and a risk of knowledge evaporating. Traditionally hands-on roles, such as master, superintendent, etc. may be replaced by people with no sea experience. If a vessel breaks down, how can it be reached quickly? There are also other security issues such as the hijacking of autonomous vessels.

The difference between autonomous and unmanned ships must be understood. Regulations will have to be changed to account for these ships, e.g. an unmanned ship cannot legally satisfy IMO COLREGs (Convention on the International Regulations for Preventing Collisions at Sea).

Legal issues

Who is at fault in the event of a cyber attack? Who is liable: equipment manufacturer, designer, supplier, ship owner? There is also very little delineation

Table 8.2 Typical hazards associated with trends in use of ICT

Trend in industry	Associated IT hazard
Performance and fuel economy (especially on platform side)	Not working in context of broader system (not meeting operational requirements)
Legislation (environment, safety)	Poor usability
	Version control
Product evolution (future proofing, software tends to be Microsoft-based because of programmer competence and cost)	Exceeding capabilities of hardware Unexpected behaviour Unnecessary behaviour
All computer systems tend to be Intel/uSoft because of programmer competence and cost	Common mode failures Software security Integrity
Competence shortfall (STCW/recruitment/retention)	Incorrect response to failures Inability to diagnose problems
Integration/flexibility of control/optimization	Lack of deterministic behaviour, undefined limits to control.
Communication (on and offship, data sharing, update/maintenance)	Lack of deterministic behaviour Dynamic management of interfaces
Data capture and analysis (operational, service, monitoring, regulation)	Confidentiality, validity of data, currency of data (timing/stamping)
Lack of expertise with ICT	No/incorrect diagnosis of ICT defects. Lack of trust in technology. Inability to perform front line maintenance. Incorrect reporting of defects
Malicious attack on IT systems	Partial failure. System behaviour designed to damage ship/data Corruption of software and data
Autonomous systems	Organisational integration. Sensor integrity. Sensor fusion. Human–system interaction strategy. Failure strategy. Third-party interaction
Analysis of data	Validity of analysis. Action in the event of incomplete data. Identification of limits. Quality of data
Visualisation of data	Information not presented to user. Information not understood by user. Too much information for user

of product liability and maritime regulations which makes it difficult to prove who is the negligent party in a cyber attack. The duty of good seamanship cannot be discharged by an autonomous ship.

There are additional legal issues relating to data protection, authentication and access to ICT systems, particularly when personal data are exchanged.

In addition to the above general concerns and issues, some of the specific technical hazards related to trends in the use of ICT in the marine industry are detailed in Table 8.2 from an LR technical report *Cybersecurity: Class Risks* (Earthy & Smith, 2016).

Human factors of automation

There is a body of technical knowledge concerned with how to address human factors in the context of autonomy and advanced control. Automation is often seen as something that can replace humans (Boorman, 2015) with the claim that it will reduce errors and make for better efficiency and lower costs. The 'error reduction fallacy' is widely believed and seems to be persistent (Sherwood Jones & Earthy, 2016). However, while sometimes positive (expected) effects occur, or at least to some extent, unexpected or unwanted things will also happen. Compare this to the development of for example driverless trains, 'automated' retail checkouts and shore-based or remote pilotage. We see how, ironically, the ticket machine for the train requires a person to help you buy your ticket, and how people 'forget' to scan (and thus pay for) food because they have to do the work at the checkout ('It's only fair').

In the marine context, remote pilotage has been proposed, in the name of safety and efficiency. This summary does not claim to be representing the whole discussion. It is believed that it would be safer for pilots to stay on land and not have to perform a risky embarkation, and it would be more efficient to be able to 'pilot' (from shore) in all or most weather situations. Some counter-arguments include that pilots must be on board to safely manoeuvre and will bring local expertise to where it is needed – on board the ship. Marine pilots are so far resisting the development, but who knows for how long as new developments are happening at a fast pace. At the time of writing, the first remote control of a ship (tug) was safely performed in a port (with a crew on board for safety). It remains to be seen what side effects and unexpected effects will occur.

Human error is a symptom of trouble deeper inside the system (Sherwood Jones & Earthy, 2016), a point which we also raise in earlier chapters. We know that strong and silent automation leads to lack of an overview which in turn leads to mode 'errors', automation 'surprises' and disruption of teamwork. Automation adds difficult monitoring tasks, increases the need to understand the automation as well as the process under control, and adds additional high-skill planning, customization and set-up tasks. It is, essentially, clumsy automation that takes care of routine situations (leading to boredom), but which cuts out and disrupts the operators during unusual situations (Bainbridge, 1983; Lützhöft & Dekker, 2002; Sarter, Woods, & Billings, 1997; Woods, Johannesen, Cook, & Sarter, 1994; Woods & Sarter, 2000). The number of references for this phenomenon are included to show that it is common and well known, but largely unrecognized in the maritime context.

There are four 'jokers' in the human aspects of automation. A joker is an issue that does not seem resolvable by traditional design, and even a management solution appears difficult to find (Sherwood Jones & Earthy, 2016). These issues have neither been discussed, nor investigated for marine automation or navigation. The jokers are:

1   Risk compensation, where the performance gains intended for increased safety are used to corporate commercial advantage. Compensation by individuals is well known in, for example, road traffic research.
2   Automation bias, where people believe the automation when they should not, and frequent errors of omission in manual operations are replaced by major errors of commission in automated operations.
3   Moral buffering, where remoteness and/or algorithms lead people to act in an inhumane manner.
4   Affect dilemma, where people attribute personality to automation – probably inappropriately, but unavoidably. We see it in how people try to reason with automation and try to relate to it at work from a safety perspective. We already make an effort to avoid upsetting Siri and Alexa.

The safety of a staffed operational system requires a defined, engineered and maintained balance of system usability and operator competence. To achieve this, the needs of humans must be considered throughout the life cycle, i.e. the design optimizes the effectiveness, efficiency, safety, environmental impact and satisfaction of staff. Standards define and operationalize usability (ISO 9241 Part 11), accessibility (ISO 9241 Part 171) and quality in use (ISO/IEC 25010). Quality in use is: 'The degree to which a product or system can be used by specific users to meet their needs to achieve specific goals with effectiveness, efficiency, freedom from risk and satisfaction in specific contexts of use'. The approach to testing is defined in ISO/IEC 25066.

Application of the regulatory framework

This section reviews components of the regulatory framework in relation to manning, human-centred development and safety management to manage the human element issues described above for unmanned ships and ships with sufficiently high levels of ICT that the relationship between the officers and ship is changed in character.

Legal aspects

Veal and Tsimplis (2017) analyse the regulatory requirements for 'unmanned' ships with varying degrees of autonomy. Their findings are summarized into a table relating each section to considerations for safety management. Table 8.3 summarizes their tests for each of the major families of regulatory requirements for protection of human life and the marine environment.

Veal and Tsimplis believe that unmanned ships of conventional design engaged on traditional trades are likely to have the same navigational rights and limitations of liability afforded to manned ships, if they can comply with the regulatory framework for manned ships. The established regulatory framework is to some degree flexible in its interpretation as to how human oversight of the ship is achieved, thus allowing a range of technologies to be used, but it clearly requires

Table 8.3 Unmanned ships' compliance with existing international regulatory framework

Requirements	Testing	Conclusion
Construction, equipment and seaworthiness	Unmanned operability requires new infrastructure, communications systems, sensors and assurance framework	• Unmanned ships should be shown to be as safe as conventional manned ships. • Unmanned ships should comply with higher standards than their manned equivalents, as the inability to deal with small incidents would create risks for others on the sea, the environment and their own cargo. • Unmanned ships should be salvor-friendly to protect the marine environment.
Manning of ships and training of crews	Who is the master?	• A master does not need to be on board, and could be a remote controller. • The issue of criminal liability needs to be resolved so that unmanned ships may operate internationally. • The liability could be assigned to a representative of the unmanned shipping company in each country, in addition to the shore-based 'master' and the owner. • For fully autonomous unmanned ships, any shore-based individual may be eligible for the role if he or she gives the command to programme the ship for a particular voyage. • The role may not be deferred completely to the automated computer system or artificial intelligence in a fully autonomous ship.
	Crewing requirements	• National maritime authorities need to be persuaded as to the reliability of the communications systems. • No regulatory guidance exists regarding how such operations can be conducted safely.
	Training and qualifications	• Unmanned ship remote controllers need training to ensure that the navigation of unmanned ships is conducted in a manner comparable to manned ships. • In addition, all staff need training in communications and software technology.
Collision avoidance, signalling and communications	• Unmanned ships and seamanship • Lookout requirements • Navigational priority and signalling	• If the communications system is sufficiently instantaneous, this can be performed by a remote controller, assuming this person has the requisite training as above, including seamanship. • This necessitates permanent supervision of even fully autonomous ships by a remote controller, ensuring 24-hour visual and aural oversight by someone able to immediately assume remote control. • Fail-safe detection and indication of not-under-command status, even during communications outage.
The duty to render assistance	UNCLOS (United Nations Convention on the Law of the Sea), Art.98 requires the masters of ships to render assistance/rescue persons in distress	• The rescue capabilities would dictate how its 'master' discharges this obligation, rather than the other way around. The remote controller can relay distress information or act as a hub for communications and possibly release life-saving devices. With no personnel on board themselves, there is no quid pro quo with other sea users.

Source: Adapted from Veal and Tsimplis (2017)

human oversight and evidence that it can and will be applied. It probably does not cover complete dependence on an automated system. They conclude that 'the control method utilized by an unmanned ship ... whether it is remotely controlled or instead pre-programmed to operate autonomously, has a profound bearing on its ability to comply with the obligations of this framework, not least in respect of collision avoidance' (Veal & Tsimplis, 2017).

This differentiation seems clear, but from a human factors or safety perspective it is not. The legal perspective is that as long as there is a person in command on or off the ship it enjoys the protections and privileges of maritime law. However, a human factors perspective lets us see that there are barriers to effective command from inadequate usability (of things and procedures), or indeed manning, which will effectively prevent this command being exercised. The job of the owner of a ship using advanced ICT is to ensure that there is a person in charge. This may be harder than designers and engineers think, because people are human. We have to use a human-centred approach to identify the threats to control and how to address them. The remainder of this chapter examines the use of regulatory and assessment tools for this purpose.

Preparing a safe manning submission

IMO Resolution A.1047 Principles of Minimum Safe Manning (IMO, 2011) provides a framework for the determination of safe manning levels for a ship. A.1047 Annex 5 explains that the minimum safe manning document submitted by the company to the administration should include a determination of the specific personal qualifications, operational policy and procedures, and infrastructure/ technology necessary to perform operational functions. The effect of 'capability enabling technology' (e.g. automation, autonomy and remote operation of functions or the whole ship) needs to be considered in the submission. Table 8.4 reviews the principles, guidance and framework for application of Minimum Safe Manning (purpose: ships that are sufficiently, effectively and efficiently manned) for issues related to ships which are unmanned or differently manned using automation and complex ICT.

The determination should consider each of the following scenarios:

1   Normal operations using the provided information, communication, control and situational awareness systems over the full duration of a voyage.
2   Situations requiring high manning in one or more departments (not only navigation) for immediate and extended periods.
3   Abnormal or novel operations requiring specialist operational knowledge, such as extreme weather, support of ships in distress, taking on riding crews.
4   Failure or damage requiring specialist advice and interaction with a range of services, such as diagnostics, use of redundancy, emergency routing, call for assistance.

Table 8.4 Application of MSM principles, guidance and determination for advanced ICT and unmanned ships

Requirement	IT issues	Manning issues
Factors to consider	Understanding of the scope and dependability of the provided ICT	Considerations regarding remote operation or support
Level of ship automation	Actual capabilities of automation. Degree of integration, reliability and availability of services provided.	Amount of 'integration work' required by crew. Proportion of time that systems can be unavailable. Degree of 'in the loop' remote control required vs supervision 'on the loop' or executive instruction (autonomous). Training to use each.
Method of maintenance used	IT service management and asset management. Service level agreement. Policy for software upgrades. Type of diagnostics and interactive support. Strategy for operation between maintenance. Redundancy/ backup requirements.	Speed and scope of offship technical support. Competence in remote diagnostics and configuration. In house and outsourced remote support. Provisions for riding crews. Any additional competence required onship.
Cargo to be carried	Sensor technologies.	Use of technologies associated with remote cargo management.
Trading area(s), waters and operations in which the ship is involved	Availability and quality of communications.	Hours of operation. Use of alternative control centres. Safe access to ship for manned operations.
Extent to which training activities are conducted on board	Quality and maintenance of training software.	Training for remote operational and support staff.
Degree of shore side support provided to the ship by the company	Level of access to ship provided by ICT. Required integrity of ICT data, monitoring or remote control.	Operational and manning policy for company. Operational and manning strategy for ship. Management structure for fleet and ships. Policy for allocation of crew/staff to ships. Factors to consider in allocation.
Applicable work hour limits and/or rest requirements	Effect of use of IT on fatigue.	Additional workload associated with IT systems. Additional reporting requirements. Workload policy for offship staff. Relationship to regulatory requirements.

continued ...

Table 8.4 (continued)

Requirement	IT issues	Manning issues
Functions performed	*Additional crew responsibilities in relation to automation of these functions*	*Considerations arising from changed manning of these functions*
Navigation planning	Maintenance of e-navigation systems and data.	Shore-side competence in navigation. Support for per-voyage decision-making.
Navigation watchkeeping	Usability of interfaces. Situational awareness.	Competence, including sea sense. Contingency for increased workload and complex situations.
Manoeuvring	Usability of interfaces (feel, latency, situational awareness). Local communications.	Monitoring (avoiding single person error). Ship handling competence. Contingency for complex situations.
Cargo handling and stowage	Electronic reporting. Supply chain/port/ship interface.	Remote supervision of port, stevedores. Safety and security arrangements. Data integrity.
Operation of ship and care for persons on board	Usability of interfaces.	Provision of care, safety and security for remaining manning.
Operating, monitoring and evaluating machinery	Use of diagnostics.	Shore-side competence in engineering. Support for decision-making.
Engineering watchkeeping	Usability of interfaces.	Contingency for increased workload and complex situations.
Fuel and ballast operations	Usability of interfaces.	Shore-side competence in operations for specific ships. Contingency for increased workload.
Safety of equipment, systems and services	Strategies for control and decision-making depending on the character and extent of automation provided. Usability of surveillance and diagnostics.	Shore-side competence in operations for specific ships. Contingency for increased workload.
Operation of ships electrical and electronic equipment	Usability of interfaces.	Shore-side competence in engineering. Support for decision-making.
Safety of ships electrical and electronic equipment	Skills required to diagnose and maintain operational and information technologies from the range of manufactures for the particular ship and systems. Usability of surveillance and diagnostics.	Shore-side competence in operations for specific ships. Contingency for increased workload.

Requirement	IT issues	Manning issues
Radio communications	Digital communications and networks including maintenance of cybersecurity.	Competence in communications issues, bandwidth management and networks. Contingency for increased workload.
		On-ship, restoration of radio operator would address problems arising from increased information and reporting.
Radio watch	Facilities for relay	Alternative means of meeting ITU (The International Telecommunication Union) and SOLAS (The International Convention for the Safety of Life at Sea) responsibilities. Local radio communication.
Emergency radio services	Relayed or can't be done.	Alternative means of meeting responsibilities.
Maintenance and repair	Skills required for diagnosing and maintaining advanced communications technologies from the range of manufactures for the particular ship and systems.	Competence in communications technologies. Contingency for increased workload.
Senior watchkeeping	Administration of ERP and management information systems.	Allocation and management of senior staff responsibilities. Relationship between command and management. Management allocated to ship.
Coordination of safety, security and environmental protection activities	Usability of decision support systems. Access to the ship and systems by maintainers/salvors.	Competence in use of information resources. Competence in use of ship and staff resources. Contingency for increased workload.
Responsibility of companies	*Additional organisational responsibilities in relation to ICT provided*	*Organizational responsibilities in relation to alternative manning*
Assessment of tasks, duties and responsibilities for dealing with emergency situations	Potential for error in use of ICT (particularly consequences of single person errors), how it will be identified and what remedial actions are required. Design of automation and advisory systems to address reasonably foreseeable abnormal situations.	Policy for loss of contact. Checking for correct performance of ICT. Strategies and procedures for failure of ICT. Hazard identification and definition of response. Alternatives to onship staff in emergencies. Capability and use of ship in marine emergencies.

continued ...

Table 8.4 (continued)

Requirement	IT issues	Manning issues
New proposal in case of changes that may affect safe manning	Changes to IT may be made specifically to change operational responsibilities, or may have consequential effects on workload, responsibility and competence. Operational changes may not be compatible with ICT and create additional or more error-prone work.	Monitoring of operational performance and suitability of organisational structure for offship support. Assessing impact of changes to ICT on manning.
Approval by administration	*Administration responsibilities in relation to ICT*	*Administration responsibilities in relation to alternative manning*
Take full account of existing instruments	Availability of expertise to interpret regulatory instruments with respect to use of advanced ICT.	Availability of expertise to interpret regulatory instruments with respect to unmanned ships.
Only approve if fully satisfied	Availability of expertise to assess the claims regarding ICT and its safe use in a proposal.	Availability of expertise to assess the claims regarding offship manning and its safe use in a proposal.
MSM determination framework	*Interdependencies and interactions of operational elements for ICT*	*Interdependencies and interactions for alternative manning*
Grades/capacities of personnel to be carried together with any special conditions	Specific IT skills, training and qualifications for tasks using the equipment and systems provided (including support and repair).	Remote manning structure for operation and support. Equivalence to manned ship.
Limitations as to validity of document	Provision and correct operation of specified infrastructure and IT necessary for each task.	Loss of communication. Failure of systems on ship. Failure of shore support facilities. Adequacy of shore staffing for fleet.
Task capability determination	Consider information demands in addition to the traditional list of task demands.	Offship operational tasks. Factors that affect offship task performance. Required competence and support.

Guidance on ICT development

IMO Circ. 1512 (IMO, 2015) is the latest and most applicable IMO guidance for future developments in marine ICT. Although it was developed for electronic navigation systems the guidance that it provides is widely applicable. Circ. 1512 and the standards that it cites provide the necessary means to develop any usable software of

Table 8.5 E-navigation quality attributes and guidelines

Quality attribute	Design quality guidance	Evaluation guidance
Product and data quality	Software product quality based on ISO/IEC 25000	Quality evaluation based on ISO/IEC 2504n
Product and data quality	Software process quality based on ISO/IEC 12207	Software assurance based on ISO/IEC 15026
Meet user needs	Human-centred design process based on ISO 9241-210	Usability testing methods based on ISO/TR 16982
Security	Based on ISO/IEC 27000	
Functional safety	Based on IEC 61508	

Source: after Circ. 1512.

acceptable quality, and to manage the support of this ICT throughout its life. Table 8.5 is based on information in Circ. 1512 and shows the attributes of quality addressed by the processes described in the Circular and the standards that apply for each attribute. For embedded software in essential engineering systems Circ. 1512 recommends the use of IEC 61508 Functional Safety and for information security ISO/IEC 27000. For bridge equipment IEC 62288 provides a starting point for user interface design. Of special note is that Circ. 1512 contains a requirement for systematic management of e-navigation system quality, using a quality management system.

Circ. 1512 explains how the principles of human-centred design described in Chapter 7 and ISO 9241-210 (International Organization for Standardization, 2010) are applied to marine systems. This is a particularly important factor for cyber-enabled ships, not only for safety of low-manned or remote operations but also to encourage crew retention – as the crew is likely to have had special training in operating the ship. The principles are:

•   Operational concepts are described to match the actual context of use.
•   Input from the crew is early, continuing and effective.
•   Improvement is continuous, and captures lessons learned from experience, trials or prototypes.
•   Systems are matched to people and tasks, not vice versa.
•   Multi-disciplinary teamwork is used to design the 'user experience' of new technology and systems.

From a software quality perspective Circ. 1512 and related guidance from other sources gives us tools to check that IT is being developed and maintained to a sufficient quality. Given the safety and business consequences of IT failures with highly automated ships, good design is only the starting point and the management of IT support should be addressed as part of the company and ship safety management. The relevant standard for IT service management is ISO/IEC 20000. Marine-specific ICT and physical security are being addressed by IMO and IACS (The International Association of Classification Societies) (see below). Regulation is slowly catching up with the need to provide remote updates to software and data. The following section explores the management of ICT safety with ISM in more detail.

Operational safety of ICT within ISM

Ships are required to have international convention certificates issued under SOLAS, which includes ISM. ISM requires consideration of risks associated with the sudden failure of critical equipment. This could include software-related failures. Although there is no explicit provision to look at software in ISM, clause 10.3 is about promoting the reliability of equipment defined as 'critical equipment'. Equipment and systems in this category increasingly have a software component. IMO already defines and addresses navigation, and to some degree regulation-required software risks.

However, it is the supplier who develops software and makes any changes. Can they be assessed under ISM? Alternatively, Class could have a requirement that supports engagement with suppliers 'to satisfy Class that systems are safe'. From the point of view of data and security, Class needs information about a supplier's level of access to their systems. Class will know that the service support for ICT needs to be assessed, but not all societies address this the same way.

The company that develops software or complex software-intensive systems needs to know that an auditor is going to assess software and needs to know what to do about it. This will almost certainly require the owner/operators' buy-in. It is easy for an auditor to ask the question, but what should be done with the answers?

Auditors could make a distinction between remotely accessible systems and those that are not. Auditors could ask what does software do and what could it do (delivery and audit). The ISM fall-back is to revert to basic operations (manual). This may not be a valid mode for lower manning or more complex systems.

ISM references a documented management system (i.e. paper) and auditing is carried out in this way. This documentation would need to be extended for ICT. This is already done in auditing in other industries, especially where software is concerned. This presents an opportunity to define the link between systems and the ship and the scope and type of system.

There is a need to revisit the ISM code and its application in this area. Explicit extension of ISM to include software/cyber-safety management would provide much of the required control. MSC.1/Circ.1526 (IMO, 2016) is a good start. The requirements of ISO/IEC 20000 IT service management and ISO/IEC 19770 software asset management could be used to give a full set of management system tools to support this. Having a service agreement for electronic charts has already been addressed as part of the adoption of ECDIS. The International Association of Classification Societies (IACS) has a Cyber Systems Panel considering requirements for essential engineering systems.

Human factors assurance for advanced control and automation

This section presents a proposed form of assessment for the safety-related ergonomics aspects of remote control and automation. The EU SARUMS project proposes five methods of control for UnManned Vessels (UMV) (Örnfelt, 2016, p. 7) that are closely related to Sheridan & Verplank's ten levels of decision-making (Sheridan & Verplank, 1978). Each method describes a different relationship between the

Table 8.6 Human factors consideration for SARUMS methods of control

Method	Human Factors Consideration for Safety	Notes
*5. Autonomous/*The UMV will sense environment, define actions, decide and act. On-board system invokes functions without informing the operator.	In addition to method 4 ensure that: • strategic information is presented in an understandable form to support strategic awareness for safety decisions. • operators/senior officers are provided with tools to bring the ship/system to a safe state. And provide evidence that responsible staff can: • take control in the event of dangerous behaviour. • take control in the event of shutdown or failure. • gain safe access to ship/ systems.	At this level, the only intervention by an operator (or senior officer) is to shut down the automation and revert to a 'safe' mode of operation.
4. Monitored/Reportive (UMV reports action). On-board system invokes functions without waiting for (or expecting) a reaction from the operator.	Class/regulatory requirements for control station ergonomics. In addition ensure that strategic information is presented in an understandable form to support strategic awareness for supervisory/management decisions by systems within the scope of the notation. Note Strategic information includes performance of the control system.	At this level the operator is monitoring the actions taken by the automated system. It can sometimes be difficult to determine what control actions the system has performed during abnormal operations.
3. Delegated/ Declarational (UMV declares intention) or Management by Exception. Authority to invoke functions is transferred to on-board system. The operator has the option to object (veto) intentions declared by the UMV during a certain time.	Class/regulatory requirements for control station ergonomics. In addition, ensure that the categorisation of alerts and presentation of information in an understandable form supports full situational awareness for control of all automatic systems within the scope of notation.	Full implementation of existing Class requirement should be adequate to ensure safe operation of such systems (e.g. LR 2016b).

Method	Human Factors Consideration for Safety	Notes
2. Directed/Permissive (UMV suggests/ asks for permission) or Management by Consent. UMV has degree of on-board cognitive capability and suggests one or several actions. The authority to make decisions is with the operator.	Class/regulatory requirements for control station ergonomics. In addition, ensure that offship staff (e.g. back office staff, VTM (Vessel Traffic Management), or programmers) making judgements about operational safety: (a) have equivalent competence to make these decisions as the human operator on the ship and (b) have access to the same level of information about the ship status as the human operator on the ship.	Provisos are added: (a) because of increasing cultural pressure from ashore for the ship to comply with direct instruction (b) when advice is provided by a computer system there is a tendency for less experienced officers to not question computer output. If these provisos are not the case, and the exclusion is not accepted by the client, the assessor would need to see evidence of how the additional risks are addressed for each offship decision support system.
1. Operated (Remote control, Tele-operation or Manual Operation). Cognitive functionality is within the human operator. The operator makes all decisions, directs and controls all vehicle and mission functions.	Class/regulatory requirements for control station ergonomics.	The purpose is to ensure control by competent crew.

operators and the ship or system. The functional, performance and verification requirements for human factors will be different for each method. The changes in requirements are cumulative. It is assumed that competent crew will be in command up to method 3. At method 4 and beyond, additional human–system issues need to be considered and addressed because of changed responsibilities and manning. Table 8.6 lists the human factors considerations for safety at each of the SARUMS methods (Lloyd's Register, Cyber-Enabled Ship Notation. version 2, in press).

The generic user and system interface and development requirements for safe remote control of automation resulting from these considerations are detailed in the following sections.

Functional requirements up to method 3

1   The information presented to operators (on-board and shore-side) needs to be rationalized and presented in a suitable format and at a rate that allows crew

to perform their duties efficiently and effectively and to maintain situational awareness.

2   Alarms and warnings are to be categorized according to the urgency and type of response required by the crew. The assignment of a category to each alert is to be evaluated on the basis not only of the machinery or equipment being monitored, but also the complete installation.

These requirements would be satisfied by compliance to the relevant requirements of Class Rules (e.g. Lloyd's Register Ships (Lloyd's Register, 2016c), part 6 chapter 1 sections 1.2, 2.3 and 3 within the context of the provisos stated in Table 8.6).

Additional process requirements for method 4 and above

1   Design of equipment and systems needs to take into account the changed expectations placed on the users to operate the systems, maintain security and diagnose failures and other problems.

2   For novel systems and at method 4 and above, the development process should apply the human-centred design activities defined in, for example, IMO Circ. 1512 and elaborated in Lloyd's Register's human-centred design guides (Lloyd's Register, 2014a, 2014b) sufficiently to mitigate identified risks to safety arising from insufficient usability.

Operational requirements, particularly at method 4 and above

1   The jobs of seafarers and shore staff need to be assessed to take account of new or changed responsibilities, including security, support and maintenance of software-intensive systems. This should include an account of the role competencies (skills, knowledge, behaviours) required to fulfil the new or changed responsibilities.

2   The seafarer needs to be trained (based upon the competencies associated with the identified responsibility requirements) to understand what it means to work on or with a cyber-enabled ship. The human is the front line of cyber security.

3   The operation of the ship needs to be monitored to ensure that the human component of the smart systems is performing safely, securely and effectively and that adequate maintenance is being carried out.

4   The cumulative effect of all changes on the safe and effective performance of seafarers and shore staff needs to be considered in terms of performance, situational awareness and training.

The operational requirements are part of the context of use of the automation and should be considered as part of the human-centred design activities, including risk assessments. In addition, the ship operator should ensure that crew are assessed as competent according to their responsibilities before the ship or system is put into operation.

Performance requirements

The design of user interfaces should enhance the usability of cyber-enabled ships and systems, reduce human error, enhance situational awareness, and support safe and effective monitoring and control under normal and foreseeably abnormal modes of operation:

1   Alarms and warnings associated with cyber-enabled systems are to be categorized according to the urgency and type of response required by the operator. (This categorization is described in the IMO Code on Alerts and Indicators, A.1021, 2009.)
2   Where the facility to provide messages in association with alerts exists, messages accompanying alerts are to describe the condition and indicate the intended response or type of response required by the operator. Additionally, messages of different categories are to be clearly distinguishable from each other.
3   There should be implementation of the principles and activities of human-centred design sufficient to identify and mitigate risks arising from the operator's use or reasonably foreseeable misuse of the cyber-enabled ship or system.

Conclusion and recommendations

We conclude that use of advanced ICT will continue to increase in the marine sector for a number of reasons, not all to do with financial imperatives, and hence the management of the safety of this technology will have an impact on all organizations in the marine industry whether they choose to lead or lag in its application. On a positive note the tools exist to manage the human–system risks associated with this technology and many are described in this chapter. We hazard the following recommendations in their application, whilst noting that this is a fast-moving topic and that in this chapter we focus on the human-related aspects of this topic.

Concept of operations

This includes a clear understanding of the type of benefit expected from use of advanced ICT. The concept of operations should include a description of the context of use and resulting user needs, including what data and knowledge are required. The context of use should include 'latent errors', the dynamics of variable autonomy and virtual/remote environments. Analysis of the context of use should also address change in safe manning arrangements (onship, offship, support, competence) and the use of and change in communications: for example, comprehension and coupling in relation to situational awareness, timing issues such as speed of response, latency and data rates; the type of information, the medium used and how it is presented; the contents (no longer only data and orders); routing (e.g. ship to office, ship to support, ship to ship, ship to VTMS (Vessel Traffic Management Systems)).

Risk assessment

This should include not only the risk reduction to be achieved through the use of advanced ICT but also the design for and management of the safety and security of advanced ICT and associated data. The risk associated with failures in the management of change should be assessed. Risks that cannot be addressed in design and that are part of the operational context are tracked in the safety, security and environmental management systems. This has unavoidable links to design, data and level of service.

Operational concept

There should be a clear description of the intended use of advanced ICT by all intended stakeholders. This is to be based on an allocation of function derived from a suitable 'people are better at, machines are better at' analysis and that takes account of the socio-technical aspects of changes to the working environment. Preference is to be given to a 'chatty assistant' (Harvey & Stanton, 2014) style of interaction in preference to a 'strong and silent' human replacement. The operational concept should be maintained as a living document and used to assist crew in understanding the intended system response.

Evaluation of use

This may need to address a broader range of tasks and a more nuanced interpretation of task completion. Measures of usability need to be broadened from achievement of tasks to successful working with computers. Human-centred quality, comprising usability, accessibility, user experience and avoidance of harm forms a starting point. User experience in this context is needed that includes human–system teamwork such as understanding, social norms, politeness, patience, etc.

References

Bainbridge, L. (1983). Ironies of automation. *Automatica,* 19(6), 775–9.

Berry, T., Stokes, J., & Toomey, B. (2016). The 'cyber enabled ship': What does it mean for the operators and regulators. Paper presented at the 13th International Naval Engineering Conference (INEC) and Exhibition. Bristol, 26–28 April.

Bhatti, R. (2017). The future state of freight: safer, greener, leaner. http://www.bhp.com/media-and-insights/prospects/2017/05/the-future-state-of-freight

Boorman, C. (2015). The need for human centric automation. http://devopssummit.sys-con.com/node/3293926

DNV GL (2014). The connected ship. In M. Latarche (ed.), *ShipInsight: Trusted Information on Maritime Technology and Regulation.* Leatherhead: ShipInsight. Retrieved from https://www.synecticsuk.com/images/stories/pdfs/The_Connected_Ship_Guide_14.pdf

Earthy, J., & Smith, R. (2016). *W01955935/techrep – Cybersecurity Class Risks Techncial Report.* Southampton: Lloyd's Register Group Technology Centre.

Graveson, A. (2015). Smart ships embrace change – secure employment. https://www.rina.org.uk/Smart_Ships_Embrace_change.html

Hand, M. (2017). Owners not demanding autonomous ships, want solutions to today's issues: ABS. *Seatrade Maritime News*. http://www.seatrade-maritime.com/news/europe/owners-not-demanding-autonomous-ships-want-solutions-to-today-s-issues-abs.html

Harvey, C., & Stanton, N. A. (2014). Safety in system-of-systems: Ten key challenges. *Safety Science*, 70, 358–66.

Hollnagel, E. (2014). *Safety-I and Safety–II: The Past and Future of Safety Management*. Farnham: Ashgate Publishing.

International Maritime Organization (2011). *Resolution A.1047(27) Principles of Minimum Safe Manning*. London: IMO.

International Maritime Organization. (2015). *MSC.1/Circ.1512 Guideline on Software Quality Assurance and Human-Centred Design for e-Navigation*. London: IMO.

International Maritime Organization (2016). *MSC.1/Circ.1526 Interim Guidelines on Maritime Cyber Risk Management*. London: IMO.

International Organization for Standardization (2010). *Ergonomics of Human System Interaction – Part 210: Human-Centred Design for Interactive Systems* (ISO 9241-210:2010). Geneva: International Organization for Standardization.

Lloyd's Register (2014a). *The Human-Centred Approach: A Best Practice Guide for Equipment Manufacturers*. London: Lloyd's Register.

Lloyd's Register (2014b). *The Human-Centred Approach: A Best Practice Guide for Ship Designers*. London: Lloyd's Register.

Lloyd's Register (2016a). *Cyber-Enabled Ships: Deploying Information and Communications Technology in Shipping – Lloyd's Register's Approach to Assurance* (1st edn). London: Lloyd's Register. https://issuu.com/lr_marine/docs/lr_guidance_note_cyber-enabled_ship

Lloyd's Register (2016b). *Cyber-Enabled Ships: ShipRight Procedure – Autonomous Ships* (1st edn). London: Lloyd's Register. https://issuu.com/lr_marine/docs/lr_cyber-enabled_ships_shipright_pr

Lloyd's Register (2016c). *Rules and Regulations for the Classification of Ships*. London: Lloyd's Register. http://www.lr.org/en/RulesandRegulations/ships.aspx

Lloyd's Register Foundation (2016). Foresight review on robotics and autonomous systems: Serving a safer world. http://www.lrfoundation.org.uk/news/2016/foresight-review-of-robotics-and-autonomous-systems.aspx (June 2017).

Lützhöft, M. H., & Dekker, S. W. A. (2002). On your watch: Automation on the bridge. *Journal of Navigation*, 55(1), 83–96.

Maritime and Coastguard Agency (2008). *Human Element Assessment Tool*. Southampton: Maritime and Coastguard Agency. https://www.gov.uk/government/collections/seafarer-health-and-safety-human-element-assessment-tools

Örnfelt, M. (2016) Safety and regulations for unmanned maritime systems. Unmanned Surface Vessel Regulation Conference, Southampton 16–17 November. Retrieved from http://www.ukmarinealliance.co.uk/sites/default/files/SARUMS-2016-MASRWG.pdf

Quick, G. (2016).. Would autonomous ships be good for society? *The Maritime Executive*. Retrieved from http://maritime-executive.com/editorials/would-autonomous-ships-be-good-for-society

Reason, J. T. (2008). *The Human Contribution: Unsafe Acts, Accidents and Heroic Recoveries*. Farnham: Ashgate Publishing.

Rolls-Royce (2016). Rolls-Royce reveals future shore control centre. https://www.rolls-royce.com/media/press-releases/yr-2016/pr-2016-03-22-rr-reveals-future-shore-control-centre.aspx

Sarter, N. B., Woods, D. D., & Billings, C. E. (1997). Automation surprises. In G. Salvendy (ed.), *Handbook of Human Factors/Ergonomics* (2nd edn, pp. 1926–43). New York: Wiley.

Sheridan, T. B., & Verplank, W. L. (1978). *Human and Computer Control of Undersea Teleoperators*. Man-Machine Systems Laboratory Report. Cambridge, MA: MIT.

Sherwood Jones, B., & Earthy, J. (2016). Using human-centred design to mitigate the risk of human error in software intensive systems. *Bulletin of the Safety Critical Systems Club,* 11–16.

TRANSAS (2016). Transas unveils the future of maritime operations. http://www.transas.com/transas-unveils-the-future-of-maritime-operations

Veal, R., & Tsimplis, M. (2017). The integration of unmanned ships into the lex maritima. *Lloyd's Maritime and Commercial Law Quarterly,* May, 303–35.

Wärtsilä Corporation (2017). One Sea – Autonomous Maritime Ecosystem introduced roadmaps to autonomous shipping. https://www.wartsila.com/media/news/22-05-2017-one-sea-autonomous-maritime-ecosystem-introduced-roadmaps-to-autonomous-shipping

Woods, D. D., & Sarter, N. B. (2000). Learning from automation surprises and 'going sour' accidents. In N. B. Sarter & R. Amalberti (eds), *Cognitive Engineering in the Aviation Domain* (pp. 327–354). Mahwah, NJ: Lawrence Erlbaum Associates.

Woods, D. D., Johannesen, L. J., Cook, R. I., & Sarter, N. B. (1994). *Behind Human Error: Cognitive Systems, Computers, and Hindsight*. CSERIAC SOAR 94-01. Dayton: OH: Wright-Patterson Air Force Base.

Epilogue

When we started working on this book, we believed there was room for 'one more safety book'. The time we spent writing it has shown us that there is more to learn, and hopefully you have found that to be true as well.

It is often said that shipping is 'behind' other domains, for example in regulation. We now know that shipping was an early initiator for SMS (Safety Management Systems) but subsequently not an early adopter as seen in Chapter 3. There have been positive effects from the introduction of the ISM (International Safety Management) Code and safety management systems in companies operating ships. However, there seem to be limitations with current safety management systems based on the predominant safety paradigm – Safety-I. This paradigm presents limitations both with respect to learning in organizations and innovation that can enhance safety. The shipping industry may benefit, at least safety-wise, from building safety management systems based on both the Safety-I and the Safety-II paradigm, allowing more flexibility and exploration of opportunities by accepting increased uncertainty. One critical factor for this to be successful is probably a change in the manning of ships to allow for flexibility and exploration.

The manning/crewing structure needs to change across the industry. However, focusing only on changing the manning numbers is unlikely to help. Some operators have already put in place a radio/information officer to perform administration work and manage communications. Other companies carry an administrator from the shore office, less for IT and more for paperwork and reporting, but we may well need an IT officer, in addition to the electro-technical officer (ETO). Interestingly, the IMarEST Human Element working group is starting to challenge the addition of staff just as we are seeing oil majors starting to add staff. Their argument, explained by leader Vaughan Pomeroy, is that jobs based on old ship designs and work are probably inefficient and we should use ergonomics and socio-technical systems approaches to design suitable, effective jobs that will save manning. It is a similar argument to that used by the Chief of US Naval Operations about 15 years ago and it is hard for human factors people to disagree.

Notwithstanding the number and skills of crew, there is a weakness in the development of safety management systems when there is a limited involvement of crew members. This calls for safety management systems that facilitate such communication. We know that strict procedures limit the flexibility of work –

work as imagined vs work as done – and we also know that the value of the human contribution may well outweigh human variability or the propensity for failure; again, job and tool design is an underutilized way of reducing risk. Risk and safety perception is a multidimensional construct and can influence safety behaviour; safety awareness is invaluable but needs organizational support.

The heterogeneous characteristic of the shipping industry is another challenge. Without a concerted external pressure from flag states or recognized organizations, there is further limited input into the system. Finally, the safety management systems also need the capacity to gather and make use of information from successful operations, taking us full circle back to Safety-II.

A central theme in this book is that we take a systemic view of the complexity that is shipping and throughout we critically discuss the role of 'human error'. The label is too often applied to problems on all levels, but for all our technical, legal and training interventions, we see little effect. An interesting point to note is that marine insurance covers crew negligence, but not crew incompetence. This point provides some protection against shipping companies claiming crew negligence in all P&I (Protection and Indemnity) claims. The reader is left to speculate on the extent to which this influences the attribution of error, because in accident investigations, you will find what you are looking for. This takes us back to the definition of human error. It is an easy label to attach, but relatively difficult to address. The industry has tried for years with technology, regulation and training, and more training. We may need to find out more about the benefit of varying the types of training and consider new approaches, perhaps in real-world settings.

Given that companies are insured for crew negligence, is there any drive to change the label of human error, except by human factors sticklers? It is unfortunate that remote control and autonomy are being touted as a way of reducing human error. This is based on an assumption that simply by deducting the human from the sharp end of the system, safety should improve, but is this really the future? There is a strong engineering push, and certainly already proof of the concept. However, there are many underspecified issues such as total cost, reliability, staffing, safety management and liability. Might it be the case that the reasons for pushing for the replacement/removal of people are not financial? It is conceivable that people are an embarrassment and a threat to owner/operators. Automation obeys executive orders, does not argue back or die in embarrassing ways. In addition, automation does not need training in order to follow regulations; neither does it require rest between watches.

In this book, we have explored contemporary safety management, suggested tentative ways forward and taken the first steps towards considering safety management in remote-controlled/autonomous ships. The work in this book has provided us, and hopefully you, with some insights and some myth busting.

Index

Bold page numbers indicate tables, *italic* numbers indicate figures.